SUSTAINABLE RESOURCE USE AND ECONOMIC DYNAMICS

THE ECONOMICS OF NON-MARKET GOODS AND RESOURCES

VOLUME 10

Series Editor: Dr. Ian J. Bateman

Dr. Ian J. Bateman is Professor of Environmental Economics at the School of Environmental Sciences, University of East Anglia (UEA) and directs the research theme Innovation in Decision Support (Tools and Methods) within the Programme on Environmental Decision Making (PEDM) at the Centre for Social and Economic Research on the Global Environment (CSERGE), UEA. The PEDM is funded by the UK Economic and Social Research Council. Professor Bateman is also a member of the Centre for the Economic and Behavioural Analysis of Risk and Decision (CEBARD) at UEA and Executive Editor of Environmental and Resource Economics, an international journal published in cooperation with the European Association of Environmental and Resource Economists. (EAERE).

Aims and Scope

The volumes which comprise *The Economics of Non-Market Goods and Resources* series have been specially commissioned to bring a new perspective to the greatest economic challenge facing society in the 21st Century; the successful incorporation of non-market goods within economic decision making. Only by addressing the complexity of the underlying issues raised by such a task can society hope to redirect global economies onto paths of sustainable development. To this end the series combines and contrasts perspectives from environmental, ecological and resource economics and contains a variety of volumes which will appeal to students, researchers, and decision makers at a range of expertise levels. The series will initially address two themes, the first examining the ways in which economists assess the value of non-market goods, the second looking at approaches to the sustainable use and management of such goods. These will be supplemented with further texts examining the fundamental theoretical and applied problems raised by public good decision making.

Sustainable Resource Use and Economic Dynamics

Edited by

Lucas Bretschger
Center of Economic Research
ETH Zurich

and

Sjak Smulders
Department of Economics and ISEEE
University of Calgary
Department of Economics
Tilburg University

Springer

A C.I.P. Catalogue record for this book is available from the Library of Congress.

ISBN 978-1-4020-6292-6 (HB)
ISBN 978-1-4020-6293-3 (e-book)

Published by Springer,
P.O. Box 17, 3300 AA Dordrecht, The Netherlands.

www.springer.com

Printed on acid-free paper

Preface

This volume includes a selection of papers presented at the conference "Sustainable Resource Use and Economic Dynamics" (SURED), held on Monte Verità in Ascona, Switzerland, in June 2004. Thirty years after the publication of the famous symposium issue of the Review of Economic Studies in 1974, which started the neoclassical literature on growth theory and resource economics. The conference sought to reinforce research efforts in order to provide adequate solutions for today's challenges in the field of sustainable development. The conference compiled innovative research from resource, energy and environmental economics, and dynamic economic theory. By bringing together leading experts, junior and senior scholars in these fields, it covered a broad range of aspects regarding the relationship between natural resource use and long-term economic development.

The SURED conference made use of the wonderful surroundings on the "mountain of truth" and the remarkable history of the conference centre, which was shaped by the desire to return to a natural way of life. In this tradition, the conference aimed at finding ways of living in an economically developed world and at the same time taking into account the natural environment with its restrictions and requirements. We take the opportunity to thank the staff of the Monte Verità centre for the hospitality and the excellent service.

Chapters 2, 3, 6, 8 and 9 are reprinted from Environmental and Resource Economics 36/1 (January 2007). For their efforts to realize this volume, we would like to thank Fabio de Castro, Esther Verdries and Ian Bateman, and the Springer publishing team; the referees and discussants; and authors of all the chapters.

List of Contributors

Jeroen C.J.M. van den Bergh
Vrije Universiteit
Amsterdam
The Netherlands

Lucas Bretschger
ETH Zurich
Switzerland

Maria A. Cunha-e-Sá
Universidade Nova
Lisbon
Portugal

Hannes Egli
ETH Zurich
Switzerland

Henri L.F. de Groot
Vrije Universiteit
Amsterdam and CPB Netherlands
Bureau for Economic
Policy Analysis
The Hague
The Netherlands

Christian Groth
University of Copenhagen
Denmark

Rob Hart
Swedish Land University
Uppsala
Sweden

John M. Hartwick
Queen's University
Kingston
Canada

Tobias H. Kronenberg
Research Centre Juelich
Germany

Peter Mulder
Ministry of Planning
and Development
Maputo
Mozambique

Joëlle Noailly
CPB Netherlands Bureau
for Economic Policy Analysis
The Hague
The Netherlands

Ana B. Reis
Universidade Nova
Lisbon
Portugal

Sjak Smulders
University of Calgary
Canada and Tilburg University
The Netherlands

Susanne Soretz
University of Hannover
Germany

Thomas M. Steger
ETH Zurich
Switzerland

Cees A. Withagen
Tilburg University
and Vrije Universiteit
Amsterdam
The Netherlands

Adriaan van Zon
Maastricht University
The Netherlands

TABLE OF CONTENTS

1. Introduction to Sustainable Resource Use and Economic Dynamics

LUCAS BRETSCHGER* AND SJAK SMULDERS**

* *CER-ETH Center of Economic Research at ETH Zurich, ZUE F7, CH-8092 Zurich, lbretschger@ethz.ch.*
** *Department of Economics and Institute for Sustainable Energy, Environment and Economy, University of Calgary; and Department of Economics, Tilburg University, P.O. Box 90153, 5000 LE Tilburg, The Netherlands, j.a.smulders@uvt.nl.*

1. Introduction

There are many compelling reasons why environmental and resource problems should be placed in a dynamic perspective. Traditionally, resource economics needs to study the dynamics of depletion of natural resources and environmental services. Current use of non-renewables, such as oil reserves, determines future resource availability. Renewable natural resources regenerate in a dynamic ecological process, which is disturbed by commercial harvesting activities. Similarly, environmental economics has to deal with pollution dynamics when pollution entails long-lasting cumulative effects in soil and marine resources or in the atmosphere. Looking at the impact of resource scarcity and pollution for the economy as a whole we additionally find that *macroeconomic dynamics* become highly relevant. To offset the increasing scarcity of natural resources and to promote sustainable development, capital accumulation and technological change are essential. In particular, the development and adoption of new technologies allow improving resource and abatement efficiency. Finally, *social dynamics* are important: the behaviour of polluters or natural resource users, as well as policy-makers, changes over time because of learning behaviour, or because of changing perceptions, the building-up of new information, and the reaction thereupon.

The experience of the world economy with oil prices over the past few decades illustrates some of the interactions between resource dynamics and macroeconomic dynamics. The present situation shows similarities with the 1970s and 1980s, when oil prices rose sharply and pollution issues entered the political agenda. In the last four years, the increase in oil prices was similar in scale to the price jumps of 1973/74, 1978/80, and 1989/90, all of which were followed by worldwide recession and rising inflation. However, historical parallels have to be handled with great care. The big recession of the mid-1970s was not only due to oil shortages but was additionally caused by other factors like the breakdown of the Bretton-Woods currency system and a broad uncertainty about the growth perspectives in general. Also, in the recent past, price increases of raw materials have been more gradual, giving households and firms more time to adjust. The

L. Bretschger and S. Smulders (eds.), Sustainable Resource Use and Economic Dynamics, 1–16.

most important difference to thirty years ago, however, is that developed countries use half as much oil per unit of real GDP as in the mid-1970s, thanks to improved energy efficiency, a switch to alternative energy sources, and the shift from manufacturing to services.

The concern for sustainability provides another illustration of the interaction between resource dynamics on the one hand and macroeconomic and social dynamics on the other. After a long process of growing awareness and changing perceptions of links between resource use, environmental problems, poverty, and intergenerational fairness, the notion of "sustainable development" is nowadays widely accepted as a main principle for environmental and development policies. However, the concept of sustainable development as used in the policy debate and among non-economists has been far away from the traditional welfare analysis in economics. Economists have succeeded to bridge a large part of the gap by taking explicitly into account natural resource constraints on output, studying resource markets, acknowledging externalities in resource use, and considering alternative ethical foundations for welfare functions and discounting principles. Accordingly, a large part of the formal literature on sustainability studies how utility levels can be sustained in a model world with (non-)renewable resources. Substitution has become the core of economists' view on sustainability. Over time, decreasing per capita amounts of natural inputs have to be sufficiently compensated for by the accumulation of man-made inputs. The greater the saving effort of the present generation is, the more feasible becomes the substitution of natural resources in production and consumption. The key research question for economists is to determine the returns and incentives of these sustainability-enhancing investment activities.

The dynamics of technological change cannot be ignored in this context: both as a threat to sustainability (in the guise of resource-using or energy-using technological change) or as the solution (more efficient resource use, clean technologies, and backstops, i.e. resource-saving technological change). Understanding the sustainability of long-term development therefore requires insight into the pace, direction, and determinants of technical change. The new growth theory that started in the 1990s provides a modelling framework in which technological change is an endogenous variable: knowledge – embodied in new capital goods, production processes, and products – is an ultimate substitute for resource inputs, without making the latter unnecessary. Developing useful new knowledge is costly and time-consuming, which turns innovation into an economic investment problem. Theories of endogenous innovation examine the incentives for innovation in a particular direction (resource-using versus resource-saving), as well as the opportunity cost of technological change resulting from crowding out of conventional investment by environmentally oriented investment. Technological progress is often modelled as incremental, which leads to a steady, but possibly moderate improvement of resource efficiency. In addition, we need to look for technology options that bring about a quantum jump in the efficiency of using natural resources. Only with radical innovations will the economy be in a position to tame the increasing resource demand in the future, given the rapid economic development today, for example, in China, India, and other emerging economies.

The complexity and breadth of sustainable development requires an even broader view, as reflected in the United Nations Millennium Development Goals, where the reduction of poverty, hunger, disease, illiteracy, and discrimination against women are the most important issues. In the future, economic analysis will be increasingly devoted to local community actions, the dynamics of social norms, and their impact on resource use in smaller groups.

The nine chapters following this introduction study different aspects of resource use, pollution, economic dynamics, and sustainability. Chapters 2–6 study the incentives to invest in clean technologies. Chapters 7 and 8 turn to substitution between energy or polluting inputs on the one hand and man-made inputs on the other. Of these chapters, Chapter 7 explores long-run growth and non-renewable resources and – together with the remainder of this introduction – provides an overview of the main modelling issues and insights from recent dynamic resource and environment modelling. Chapter 8 turns to the empirics of gradual improvement and international convergence in aggregate energy efficiency. Chapter 9 focuses on local communities and how social norms with respect to resource use evolve there. Chapter 10 uses insights from economics to reinterpret classical mechanics. In the remainder of this chapter we discuss the main common elements and a unifying modelling framework for the chapters in this volume.[1]

2. Growth and Pollution

2.1. POLLUTION IN DYNAMIC THEORY

Growth and pollution have been studied extensively over the last decade, both empirically and theoretically. From an empirical perspective, the Environmental Kuznets Curve (EKC) hypothesis has been most visible, although most of the earlier work under this heading looks into the relationship between *levels* of income and pollution. Only recently has growth been explicitly studied (Bradford et al. 2005; Brock and Taylor 2004). The theoretical analysis builds on the "endogenous growth" literature developed in macroeconomics (starting with Romer 1986, see Aghion and Howitt 1998 for a broad exposition). The most elementary endogenous growth model, the "AK model", extended for basic environmental and resource aspects, provides important insights into the links between investment in production capacity and the resulting economic growth on the one hand, and the polluting consequences of production and environmental policy on the other.

There is still a big gap between the empirical and theoretical literature. The Environmental Kuznets Curve literature typically aims at characterizing the relationship between levels of income and pollution, without linking empirical model specification to theory, and without testing for underlying mechanisms. The theoretical literature normally restricts the analysis to constant growth (or balanced growth) paths and ignores the richer dynamics emerging from the empirical EKC studies in which the pollution–income link changes over time or with income levels.

The chapters by Cunha-e-Sá and Reis, Egli and Steger, and Soretz, in this volume, fill some of the gaps. They further develop the AK model to investigate the relationship between economic growth and environmental policy. In particular, they introduce new dynamic elements that allow for a more detailed study of clean technology adoption, uncertainty, and the link to the EKC. In chapters by Hart, and Van Zon and Kronenberg, the dynamic impacts of environmental policy and induced innovation are discussed more in detail.

To give a clear view on how we can start to study environmental economic dynamics from the canonical AK model, we first briefly review the AK approach and then show how clean technology can be modelled, how the pollution–income link depends on abatement technology, and how uncertainty can matter in this context.

2.2. THE ENVIRONMENTAL AK FRAMEWORK

The distinguishing feature of the AK model is that aggregate production in the economy, Y, is linearly related to a broad measure of reproducible capital, K, in the following way:

$$Y = AK. \tag{1}$$

Accordingly, the marginal product of capital is given by A; it determines the rate of return and incentives to invest. The aggregation of all relevant man-made capital goods into one stock variable that is linearly proportional to output simplifies the analysis considerably.

To incorporate environmental aspects into the AK model, pollution can be modelled as a by-product of either inputs (K) or consumption (C); abatement expenditures (E) are assumed to reduce pollution for given polluting input levels. Hence, the general formulation for the pollution generating process can be written as:

$$P = p(K, C, E) \tag{2}$$

where $p_C \geqslant 0,\ p_K \geqslant 0,\ p_E \leqslant 0$ (with the subscripts denoting first-order partial derivatives).

Pollution[2] is assumed to affect (as an externality) both production, through an effect on productivity level A, and instantaneous utility U, which otherwise depends on consumption C. Thus we can write:

$$A = a(P) \tag{3}$$

$$U = u(C, P) \tag{4}$$

where $a_P \leqslant 0,\ u_C \geqslant 0,\ u_P \leqslant 0$. Growth of output and levels of pollution are determined by the allocation of total production over consumption, capital investment, and pollution abatement. Investment in the economy (dK/dt) and investment in the environment (E) come at the cost of consumption (C), according to the following goods market equilibrium condition:

$$Y = C + E + dK/dt. \tag{5}$$

Now consider a balanced growth path along which all terms in (5) grow at the same rate so that the ratios C/Y, E/Y, and $(dK/dt)/Y$ are constant. If the pollution generating process in (2) has properties such that we can write it in the following specification:

$$P = p(K/E, C/E, 1) \tag{6}$$

pollution is constant along the balanced growth path, too.[3] Thus, with the linear production function (equation 1) and "ratio-dependent" pollution function (equation 6), "sustainable growth" is feasible: output grows at a constant rate and pollution does not increase. There are no limits to growth in this case. If preferences are of the Cobb-Douglas type, a balanced growth path is not only feasible but also optimal with discounted utility maximization (see Smulders and Gradus 1996). A specification for preferences giving this result is $U = (1-\sigma)^{-1}[C \cdot (\bar{P} - P)^{\phi}]^{1-\sigma}$, where $\bar{P}$ is the critical value of pollution beyond which welfare cannot be sustained. With additive preferences, however, e.g. $U = (1-\sigma)^{-1}C^{1-\sigma} + (\bar{P} - P)^{\phi}$, it is optimal to spend a larger and larger part of output on abatement and to invest less and less in capital accumulation so that the growth process comes to an end (Stokey 1998).

Analytically, the model defined by equations (1)–(6) is an extremely convenient specification. Only one stock variable matters, viz. K, and no transitional dynamics arise. However, the specification in equation (6) might be seen as an overly optimistic view: doubling capital, consumption, and abatement does not double pollution but in fact leaves pollution unaffected. This implicitly assumes strong learning effects or technological change that offset the "scale effect", defined as the tendency of pollution to expand with the scale of economic activity, keeping fixed the production technology and the composition of output (cf. Brock and Taylor 2005). A standard replication argument would produce a completely different result: doubling all inputs would double all outputs, like building next to a factory another identical factory would double pollution. The absence of constant returns to scale calls for an explanation in terms of increasing returns or technological change. First, when expanding the scale of the economy the productivity of abatement might increase (or the polluting consequences of capital might diminish) due to increasing returns: new firms that enter the economy bring new knowledge, broaden the scope for learning and experimenting, and might thus increase the productivity of abatement. Alternatively, over time technological change may improve the productivity of abatement or may cause pollution per unit of output to fall.

The environmental growth models by Cunha-e-Sá and Reis (Chapter 3) and Egli and Steger (Chapter 2) make the learning and technological change effects that are hidden in equation (6) more explicit. To connect these papers to the specification in equation (6), we need to disentangle the technology/productivity effect from the input effect of abatement. Capturing the former by T_P and using a simple iso-elastic specification, we specify the pollution-generating process as:

$$P = \frac{K^\eta E^{1-\eta}}{T_P} \tag{7}$$

where $\eta > 1$.[4] In this specification, doubling the rival inputs E and K doubles pollution, but improvements in the technology parameter T_P reduce pollution. To capture learning-by-abating, we assume a positive link from abatement to technology:

$$T_P = E^\gamma. \tag{8}$$

If $\gamma = 1$, equations (7) and (8) give $P = (K/E)^\eta$ which is consistent with equation (6) and thus allows for sustainable growth. This justifies the approach in older papers (e.g. Smulders and Gradus 1996) and newer ones (e.g. Soretz Chapter 6).

2.3. LEARNING-BY-ABATING

In Chapter 2, Egli and Steger open up the black box further and are more explicit about the sources of learning-by-abatement. Their parametric example of the pollution equation can be written as:

$$P = \frac{C - C^\delta E^{1-\delta} T_E}{T_P} \tag{9}$$

$$T_E = E^\gamma \tag{10}$$

where $\delta \in (0, 1)$. In equation (9), consumption, C, rather than (capital) inputs, K, is polluting and abatement E has an additive effect rather than a multiplicative effect. The consequence of the latter is that we can distinguish more productive abatement technology (reflected in increases in T_E) from cleaner production technology (reflected in increases in T_P).[5] Equation (10) links abatement technology improvements to levels of abatement and thus captures learning-by-abating. As long as $\gamma > 0$, there are increasing returns so that abatement costs fall with the level of abatement. When consumption and abatement grow at a common growth rate, pollution will first rise and then fall. To see this, we rewrite equations (9)–(10) as:

$$P = C[1 - bC^\gamma]$$
$$b = (E/C)^{\gamma+1-\delta}.$$

Now assume E and C grow at the same rate so that b is a constant. Then, for small C, P grows, but for large C, P declines. Andreoni and Levinson (2001) have shown this EKC pattern in a static model with an exogenous endowment from which consumption and abatement ($C + E$) can be financed. First, Egli and Steger demonstrate that when the specification of preferences is appropriately chosen, a corresponding AK-growth model generates a (quasi-) balanced growth path along which E/C is indeed constant and P follows the EKC pattern. Second, and more generally, incorporating the specification in equation (9) in an

AK model, the authors can show how the turning points of the EKC change with technology and preference parameters. Third, they also make explicit the role of (Marshallian) externalities and the implications for corrective taxation. For example, learning could take place on the economy-wide level so that technology T_E is determined by economy-wide abatement and individual small firms can hardly affect T_E and take the level of technology as given.

2.4. CLEAN TECHNOLOGY ADOPTION

Cunha-e-Sá and Reis (see Chapter 3) are even more explicit about the technological progress in abatement. They focus entirely on pollution reduction through changes in technology (increases in T_P) and abstract from instantaneous abatement possibilities (in terms of equation (7), they set $\eta = 1$). In particular, they assume that in order to have less pollution per unit of capital, a new technology has to be installed. Because of adjustment costs, technological change is discontinuous: at discrete times the economy adopts a cleaner technology, and at periods at which there is no switch to a new technology, pollution necessarily increases with production. Note that the cleaner technology is applicable nationwide, so that we may refer to a "general purpose technology" (as in Helpman 1998). Although the authors consider a single adoption only, a series of sequential adoptions could allow for a constant or declining trend in pollution. This would go along with a sequence of investment expenditures, which is similar to the ongoing abatement expenditures in the model with flow-abatement E only.

The chapter investigates when economies optimally choose to adopt the cleaner technology and how the change in technology affects growth in the economy. While the usual EKC literature argues that environmental policy reacts to growth in income, the reverse effect is actually also important in a general dynamic equilibrium setting. Indeed, knowing that a cleaner technology that reduces pollution per unit of capital will be available in the future, society values capital more than without adoption, which boosts investment and growth. Accordingly, the paper finds that growth of consumption and capital accelerates prior to the adoption date, while these variables grow at a constant rate in the absence of adoption.

2.5. THE PORTER HYPOTHESIS AND DIFFUSION OF CLEAN TECHNOLOGY

According to the Porter Hypothesis strict environmental regulation can induce efficiency and encourage innovations that help improve competitiveness. In the chapter by Hart (Chapter 4), the incentives for innovation are studied in this perspective. There is no separate abatement technology for firms, so that, for given level of technology, firms can reduce pollution only by reducing production. However, changing the composition of output can reduce pollution per unit of aggregate output. Firms produce intermediate goods that are imperfect substitutes in final goods production (so that, the one-factor production function

(1) is replaced by a multi-input production function). Firms are heterogeneous in the sense that they differ according to vintages; newer firms pollute less per unit of output since they embody the newest pollution-saving technology; they also produce more output per unit of input. These assumptions imply that we need to replace the production function (1) and pollution function (7) by the following expressions, respectively:

$$Y = \sum_{i=1}^{N} A_i K_i \tag{11}$$

$$P = \sum_{i=1}^{N} P_i = \sum_{i=1}^{N} \frac{1}{T_{Pi}} K_i \tag{12}$$

where K_i and P_i are capital and pollution in firm i, respectively, T_{Pi} is the firm-specific pollution coefficient, N is the number of firms that are able to produce.

Over time, the number of firms expands. Research and development (R&D) activities provide entrants with new vintages of technology, which allows them to be more productive as well as less polluting. In particular, if successful in R&D, newer firms (with higher index i) have access to technology characterized by the following:

$$T_{Pi} = T_P \cdot (\gamma_P)^i \tag{13}$$

$$A_i = A \cdot (\gamma_A)^i \tag{14}$$

where $\gamma_P > 1$ ($\gamma_A > 1$) denotes the degree to which a new firm is cleaner (more productive) than the cleanest incumbent firm. The new technology becomes available with a certain probability, which increases with the amount of labour spent in R&D.

When newer firms increase their inputs (K) at the cost of older firms, production becomes less polluting. To introduce newer vintages, entrants have to incur an upfront investment cost. Now two channels arise through which environmental policy is effective. First, tougher environmental standards shift production from old to new vintages, thus reducing pollution on impact. Second, since new vintages attract a larger part of the total market and hence make more profits, tougher environmental standards trigger faster innovation, which reduces pollution in future also. Any new cleaner technology earns more profits than without the policy and firms have bigger incentives to innovate. Since by construction innovation implies higher productivity as well as cleaner production in the newest vintages, the Porter hypothesis materializes in the model: environmental policy not only reduces pollution but also increases growth.

Similar Porter-hypothesis effects arise in the vintage model by Van Zon and Kronenberg (Chapter 5), although their model is more complex. R&D produces two different types of innovations: radical or incremental ones. Radical innovations entail new General Purpose Technologies (GPTs), which are either carbon or non-carbon based. Incremental R&D improves the productivity of these technologies by developing complementary inputs. Old vintages of GPTs may require carbon-based energy, while newer vintages may require non-carbon-based energy. A carbon tax shifts profits towards vintages that do not use carbon-based energy and promotes innovation associated with these vintages. The choice between the two types of R&D is endogenous in the model and is driven by expected profits. The model generates interesting diffusion patterns: carbon-free and other new technologies may gradually diffuse in the economy when applied research targeted at these technologies cumulates. Due to the stochastic nature of R&D, history plays a significant role. The characteristics of new technologies are unpredictable. When predominantly carbon-intensive technologies have been developed in the past, a carbon tax has a different effect on innovation and growth than when the economy happens to rely on alternative energy sources.

2.6. UNCERTAINTY AND THE VULNERABILITY EFFECT

In the benchmark model it is attractive to spend on pollution reduction because it boosts utility and productivity, cf. equations (3) and (4). In Chapter 6, Soretz adds a third reason to reduce pollution: reductions in vulnerability to shocks. She assumes *expected* aggregate production equals AK, as in equation (1), but *actual* income is subject to exogenous shocks, the effects of which are larger the poorer environmental quality is. In particular, actual output is given by:

$$Ydt = K \cdot [Adt + P^{\psi} \nu dz]$$

where dz is the stochastic variable (modeled as the increment of a Wiener process) capturing the shocks to aggregate income, and ν and ψ are parameters. The bigger $P^{\psi}\nu$, the bigger the impact of a given shock dz. Hence ψ measures the effect of pollution on vulnerability to shocks. A risk-averse society spends more on abatement to mitigate the vulnerability effect. This crowds out investment in physical capital and tends to reduce growth. However, the risk itself may at the same time increase savings for precautionary motives. Moreover, higher spending on abatement may strengthen the productivity effect (see equation (3)). Both forces tend to increase the rate of economic growth. The paper sorts out the counteracting effects and formulates implications for optimal environmental taxation.

3. Resource Use and Growth

3.1. INTRODUCING ADDITIONAL INPUTS

When adding additional inputs like natural resources and labour in the model, we can analyse at least two major issues. First, a more detailed study of pollution becomes possible. Whereas the AK model and its variants in the previous section focus on pollution from aggregate economic activity, most pollution in the real world is associated with particular inputs (e.g. chemical inputs or energy use), rather than economic output. Second, the dynamic effects of resource scarcity can be studied. Both topics are closely related. Growth and environmental degradation may be decoupled through substitution of clean for dirty inputs in production. Only substitution at the highest aggregation level was incorporated in the one-factor AK models with abatement. To illustrate this point, suppose pollution stems from production, so that $\partial p(K, C, E)/\partial C = 0$. Then "net output", or output available for consumption and investment, $Y - E = C + dK/dt$, can be written as a function of pollution P and capital stock K. In particular, inverting (2) to write $E = b(K, P)$, we can write the production structure in equations (1), (2), (3), and (5) as:

$$Y - E \quad = a(P)K - b(K, P) \equiv F(K, P) \tag{15}$$

$$F(K, P) = C + dK/dt. \tag{16}$$

In this reformulation pollution acts like an input that is a substitute for capital in the function $F(.)$, which results from the fact that by varying the amount of abatement activities, a given amount of capital produces different levels of pollution and net output (cf. Stokey 1998; Copeland and Taylor 2003). Increasing K by one per cent without increasing pollution requires increases in abatement so that net output increases by less than one per cent. Hence, substitution and diminishing returns with respect to capital show up in the above-discussed AK framework once we consider net output rather than output including abatement activities. However, it is only substitution between "generic" capital and pollution; there is no distinction between clean and dirty inputs.

To make input substitution, e.g. of clean for dirty inputs or of abundant for scarce resources, as well as diminishing returns more explicit, it is useful to directly turn to a multi-input production like the following:

$$Y = AK^{\alpha} L^{\beta} R^{\omega} \tag{17}$$

where A is total factor productivity, K capital, L labour, and R a polluting input. Now Y should be interpreted as net output available for consumption and investment, as there is no need to distinguish separate abatement activities any more. With the example of climate change and air pollution in mind, one can interpret the polluting input (R) as energy. Pollution generated is proportional to energy, $P = \pi_R R$, where π_R is the pollution content of energy (e.g. carbon content). The Cobb-Douglas specification in equation (17) implies a unitary elasticity of substitution, which is restrictive and perhaps unrealistic, but suffices to illustrate some insights that survive with lower substitution possibilities.

3.2. NON-RENEWABLE RESOURCES AND GROWTH

In Chapter 7, Groth uses the representation of production possibilities in equation (17) to study the impact of scarcity of resources on economic growth. Suppose R is the use (extraction) of a non-renewable resource, which implies that the total amount of input use from now on until the indefinite future is limited by the current stock of resources. Since no production is viable without resource use

($Y = 0$ whenever $R = 0$), production can only be sustained if in the end smaller and smaller resource flows are extracted over time. To sustain a constant or growing production level, the decline in resource use has to be offset by increases in substituting inputs or by increases in productivity of inputs through technological change.[6] Suppose that the economy is on the balanced growth path along which production and produced capital goods grow at a common rate g, so that $(dY/dt)/Y = (dK/dt)/K$, and that a constant fraction of the population is in the workforce, so that $(dL/dt)/L$ equals population growth. From (17), we can then express per capita output growth:

$$\frac{dY/dt}{Y} - \frac{dL/dt}{L} = \frac{1}{1-\alpha}\left[\frac{dA/dt}{A} - (1-\alpha-\beta)\frac{dL/dt}{L} + \omega\underbrace{\frac{dR/dt}{R}}_{<0}\right]. \quad (18)$$

Since resource use has to ultimately decrease over time ($dR/dt < 0$), balanced per capita growth can only be positive if one or more of the following holds:

1. there is technical change ($dA/dt > 0$), so that higher productivity of inputs offsets the adverse effect of declining per capita resource use on production;
2. there is no technical change, non-increasing returns, and declining population ($dA/dt = 0, \alpha + \beta + \omega \leqslant 1$ and $dL/dt < 0$) so that the declining resource stock no longer translates into lower per capita resource endowments;
3. there is no technical change, positive population growth, but (mildly) increasing returns ($dA/dt = 0, dL/dt > 0$ and $\alpha + \beta > 1$), so that declining resource use is offset by productivity increases from scale economies;
4. there is no technological change and no population growth but (strong) increasing returns to scale ($dA/dt = dL/dt = 0, \alpha > 1$), so that the accumulation of capital offsets the productivity losses from declining resource use.

More precise conditions cannot be given until we know the determinants of resource use, technological change, and population growth. Keeping population growth as an exogenous variable, Groth explores in Chapter 7 how increasing returns and endogenous technological change in various guises, can provide the economy with increasing factor productivity to offset declining resource use.

Important policy implications can be drawn from equation (18). For example, conservation policies that slow down the rate of extraction result in a higher (but still negative) value for $(dR/dt)/R$ and growth of output can be higher. In this sense promoting long-run growth and "supporting the environment" go hand in hand. Of course initial levels of output are lower with reduced resource use.

More broadly, equation (18) shows that regulation can affect the long-term growth rate only through affecting technological change, population growth or the rate of extraction. However, technological change and extraction should be considered as endogenous variables, and the important question is how these variables can be controlled separately. Groth shows that only in a special case can policy affect the rate of technological change permanently and independently of extraction. In this special case, the generation of new technology must be represented by an equation like

$$dA/dt = A \cdot g(L_A) \tag{19}$$

where L_A is the amount of labour devoted to R&D and $g(.)$ is an increasing function. The key characteristic is that a given rate of technological change, $(dA/dt)/A$, can be maintained as long as labour input in R&D is constant, as in most early endogenous growth models (Romer 1990; Grossman and Helpman 1991, see also equations (13) and (14)). Although this specification has been used in most of the literature on endogenous growth and non-renewable resources, it is non-robust and biases the conclusions in an optimistic direction. In the more general case, new technology creation requires labour (L_A), capital (K_A), resources (R_A), and existing knowledge incorporated in A:

$$dA/dt = G(A, L_A, K_A, R_A). \tag{20}$$

Hence, the specification in equation (19) is limited to constant returns to A and absence of capital and resources as inputs. Any relaxation of these knife-edge assumptions dramatically alters the possibilities to affect the long-run rate of technological change. First, when we deviate from constant returns to A, so that $dA/dt = h(A) \cdot g(L_A)$ with $h(.)$ a concave function, the long-run rate of technological change is determined by parameters that cannot be affected by policy easily (cf. Jones 1995). Second, when resource use enters the technological progress function (equation 20) either directly or indirectly because R&D requires capital and the production of capital requires resources, the rate of extraction affects the rate of technological change.[7] Groth shows how in the general or robust case no taxation of any kind has long-term effects on growth unless they affect the depletion rate, which the usual taxes (e.g. a research subsidy, an interest income tax, and an investment subsidy) do not.

3.3. INTERNATIONAL PRODUCTIVITY CONVERGENCE

Growth without deteriorating environment is likely to require substitution of clean for dirty inputs and reduction of the use of energy or other polluting inputs per unit of output. To explore how we can accomplish this and what are the implications for growth, we assume constant returns to scale and derive from equation (17) the following expressions for average productivity of capital and the amount of pollution per unit of output (which we will label the pollution intensity):

$$\frac{Y}{K} = A\left(\frac{L}{K}\right)^{\beta}\left(\frac{R}{K}\right)^{1-\alpha-\beta} \tag{21}$$

$$\frac{R}{Y} = A^{-1/(1-\alpha-\beta)}\left(\frac{Y}{L}\right)^{\beta/(1-\alpha-\beta)}\left(\frac{Y}{K}\right)^{\alpha/(1-\alpha-\beta)}. \tag{22}$$

The productivity of capital is no longer a constant A, as it was in (1), but declines with capital under the standard neoclassical assumption of diminishing returns to capital (i.e. $0 < \alpha, \beta < 1$). Due to input substitution, capital productivity increases with energy use (and hence with pollution). Furthermore, in equation (22) capital is no longer polluting, as it was in equation (7), but is in fact a clean substitute for polluting inputs in net production (cf. equation (15)). Finally, we note that technological change (increases in A) reduces the pollution intensity: it reduces inputs per unit of output and therefore reduces pollution per unit of output.

When pursuing sustainable growth, the reduction in energy use per unit of output is crucial. According to equation (22), this is possible by relying more on clean inputs, L and K, in production. The question is whether and where this is possible. We find an elementary answer if we close the model by the Solow-like assumption of a fixed savings rate (cf. Brock and Taylor 2004, for a related argument). A fixed fraction, say s, of output is assumed to be invested in capital so that capital grows at rate sY/K. Hence capital grows quickly when capital productivity Y/K is large. Note from equation (22) that a large capital productivity Y/K also implies a high pollution intensity. A fast rate of growth of capital implies that capital productivity falls over time, see equation (21), and that pollution intensity falls, see equation (22). Hence, we arrive at a *convergence result*: a high initial pollution intensity implies fast reductions in pollution intensity over time, and vice versa, low pollution intensities imply slow reductions in pollution intensity. Countries with differences in pollution intensity therefore tend to converge in terms of pollution intensity.

An alternative source of convergence in pollution intensities is technology diffusion. There exist enormous international differences in technology (total factor productivity). Poor countries not only have relatively little capital (and hence high capital productivity Y/K and high pollution intensity R/Y), but also relatively low technology levels A, which gives scope for imitation and absorption of foreign technologies, relatively fast growth in A and hence relatively fast reductions in pollution intensities.

In Chapter 8, Mulder and De Groot test the convergence hypothesis for pollution intensities within a production function framework, assuming energy is the polluting input. In doing so, they compare their results with convergence in labour productivity. They emphasize the importance of studying dynamics both at the aggregate and sectoral levels, as data aggregation to single country observations may obscure sectoral convergence. They use data from 4 main sectors and 10 subsectors in manufacturing of 14 OECD countries in the period 1970–1997. They first observe that cross-country variation of energy productivity is much higher than that of labour productivity. In addition, the authors find evidence for conditional convergence of energy and labour productivities in most but not all sectors of the economy. It is important to note that the results for β-convergence in their paper are conditional on country-specific conditions, so that absolute international productivity differences are predicted to persist in the long run. Notably, in the σ-convergence analysis energy productivities are found to diverge on a macroeconomic level, so that scale, market and policy effects within countries are confirmed to be essential for the productive use of resources.

4. Intercommunity Social Dynamics

So far we have ignored spatial aspects of resource dynamics. We have seen in the previous section that different national economies have their own specific characteristics, and international contacts might give rise to convergence or divergence over time of resource-use patterns. Geographical specialization in resource use may change resource dynamics directly. Indirectly, resource use is affected by the macroeconomic dynamics stemming from the accumulation of complementary assets and spatial diffusion of new technologies, as well as the social dynamics related to the spatial spillovers of social rules. Such rules seem to be especially important when we leave the country level and focus on the level of local communities.

Local communities may not only differ with respect to resource availability and productivity in harvesting, they may also be governed by different social norms concerning cooperation. These norms are subject to their own (social) dynamics. Studying the interaction between resource dynamics and social dynamics is rewarding in at least two respects. First, often policy is faced with a situation characterized by local communities and a spatial distribution of activities. Second, the need for policy is weakened by the capacity of some of these systems to spontaneously generate social norms. Especially in relation to natural resource use, local communities can involve local mechanisms of monitoring and control, which (partially) replace hierarchical public policy. The combination of resource dynamics and spatial structure thus is of relevance to the formulation of optimal resource policies or institutional arrangements.

In Chapter 9 by Noailly, Van den Bergh, and Withagen, agents are assumed to harvest a common pool resource. The agents, who are either cooperators, defectors, or enforcers, are located on a circle, observing the actions of their nearest neighbours only. The specific assumption is that agents can enforce common harvesting norms by punishing the defectors not harvesting in a sustainable manner. Thus the set-up allows for a rich structure of local and global interactions in the economy; the latter consist of the impact of aggregate harvesting and the overall stock of the resource on harvesting strategies of individuals. After providing theoretical results and performing extensive numerical analysis, the authors conclude that, unlike in the previous literature, the three strategies can coexist in a large variety of constellations, while cooperators are very likely to be present at all times. Furthermore the authors emphasize that, when resource dynamics are included, cooperative equilibria become even more likely.

5. Sustainable Motion in Classical Mechanics

Traditionally, formal dynamic modelling in economics has borrowed a lot from classical mechanics. In Chapter 10, Hartwick takes the opposite direction. He uses insights from economics to reinterpret classical mechanics. He observes that "particle" motion in classical mechanics has an account in energy units of current product (a two element vector) balanced with current input (a two element vector). With sustainable or periodic motion each product element is balanced over the period of motion with its own input period after period. There is no cross-subsidization. With non-periodic (non-sustainable) motion, there is cross-subsidization and the values of the inputs are not maintained. The current balance energy account derives quite directly from Hamilton's equations characterizing equilibrium motion.

6. Conclusions

We have argued that studies that combine resource dynamics with macroeconomic dynamics and/or social dynamics provide new insights into the issues of sustainability, the turning points of the Environmental Kuznets Curve, technology adoption, and induced innovation, protection against environmental disasters, long-term effects of resource scarcity, pollution intensity convergence, and local cooperative behaviour in resource extraction. We have shown an underlying and unifying framework of modelling production, pollution, and abatement for these topics. We expect future work to deal with a more detailed analysis of different types of technological progress in production and abatement technology, the role of uncertainty and radical technological change, and the microeconomic foundations of semi-reduced-form modelling of abatement. We hope that in the future the links between dynamic theoretical models and econometric time-series or panel analysis will be further strengthened.

Notes

1. This introduction partly reproduces and extends Bretschger and Smulders (2007). In particular, Sections 2.5, 3.1, 3.2, and 5 did not appear before.
2. To simplify our exposition, we assume that the *flow* of pollution determines productivity and utility. Many environmental problems, however, instead relate to the *stock* of cumulated pollution (concentration levels).
3. Let s be the savings rate $s = (dK/dt)/Y$, which is by definition constant along a balanced growth path. Then K grows at rate $(dK/dt)/K = sY/K = sA = sa(P) = sa(P(K/E, C/E, 1))$, which is a constant. Since A is a constant, Y and K grow at the same constant growth rate.
4. The iso-elastic specification has the problem that zero abatement ($E = 0$) implies infinite pollution (Brock and Taylor 2005 P.1805). Therefore, equation (7) should be interpreted to hold only for a minimum level of abatement. A similar problem arises with the iso-elastic learning function in equation (8). These undesirable properties can be easily removed by replacing equations (7) and (8) by equation (7′) $P = T_P^{-1} K \min\{1, (E/K)^{-(\eta-1)}\}$ and equation (8′) $T_P = \max\{T_0, E^{\gamma}\}$, respectively. The threshold in equation (7′) implies that with zero abatement, pollution is proportional to capital and that a minimum amount of abatement is required before abatement starts to be effective. The threshold in equation (8′) implies that learning starts only for large enough abatement levels. As long as $\gamma = 1$ and $T_0 < K < E$, we still find $P = (K/E)^{\eta}$.
5. Alternative labels are pollution-augmenting and abatement-augmenting technological change. The distinction is impossible to make in the Cobb-Douglas specification of equation (7),

exactly like labour-augmenting and capital-augmenting technological change are equivalent in Cobb-Douglas production functions.

6. All that follows holds in a similar way for an economy that reduces polluting inputs over time (rather than non-renewable resources specifically), perhaps in order to improve environmental quality over time.
7. An interesting example of both these things happening is the case in which the function G in equation (20) is the same function as the goods production function in equation (17). This case is equivalent to increasing returns in the model without changes in A, see Chapter 7, Section 4.1.

References

Aghion, P. and P. Howitt, (1998), *Endogenous Growth Theory*. Cambridge, MA: MIT Press.

Andreoni, J. and A. Levinson (2001), 'The Simple Analytics of the Environmental Kuznets Curve', *Journal of Public Economics* **80** (2), 269–286.

Bradford, D. F., R. A. Fender, S. H. Shore, and M. Wagner (2005), 'The Environmental Kuznets Curve: Exploring a Fresh Specification', *Contributions to Economic Analysis & Policy* 4(1), Article 5. http://www.bepress.com/bejeap/contributions/vol4/iss1/art5.

Bretschger, L. and S. Smulders (2007), 'Sustainable Resource Use and Economic Dynamics', *Environmental and Resource Economics* 36 (1), 1–13.

Brock, William A. and M. Scott Taylor (2004), 'The Green Solow Model', *NBER Working Paper No. 10557.*

Brock, William A. and M. Scott Taylor (2005). 'Economic Growth and the Environment: A review of theory and empirics', in Philippe Aghion and Steven N. Durlauf, eds. *Handbook of Economic Growth, Volume 1B*. Elsevier.

Copeland, B. and S. Taylor (2003), *International Trade and the Environment: Theory and Evidence*. Princeton, NJ: Princeton University Press.

Grossman, G-M. and E. Helpman (1991), Innovation and Growth in the Global Economy, Cambridge, MA: MIT Press.

Helpman, E. (1998), *General Purpose Technologies and Economic Growth*. Cambridge, MA: MIT Press.

Jones, C. I. (1995), 'R&D based Models of Economic Growth', *Journal of Political Economy* 103, 759–784.

Romer, P. (1986), 'Increasing Returns and Long Run Growth', *Journal of Political Economy* **94** (October), 1002–1037.

Romer, P. (1990), 'Endogenous Technological Change', *Journal of Political Economy*, **98**, Part II, S71–S102.

Smulders, S. and R. Gradus (1996), 'Pollution Abatement and Long-term Growth', *European Journal of Political Economy* **12**, 505–532.

Stokey, N. (1998), 'Are There Limits to Growth', *International Economic Review* **39** (1), 1–31.

2. A Dynamic Model of the Environmental Kuznets Curve: Turning Point and Public Policy

HANNES EGLI* AND THOMAS M. STEGER**

* *CER-ETH – Center of Economic Research at ETH Zurich, ZUE F 10, CH-8092, Zurich, Switzerland; Author for correspondence (phone +41-44-632-0468; fax + 41-44-632-1362 E-mail: hegli@ethz.ch)*

** *CER-ETH – Center of Economic Research at ETH Zurich, ZUE F 13, CH-8092, Zurich, Switzerland*

Abstract. We set up a simple dynamic macroeconomic model with (i) polluting consumption and a preference for a clean environment, (ii) increasing returns in abatement giving rise to an EKC and (iii) sustained growth resulting from a linear final-output technology. There are two sorts of market failures caused by external effects associated with consumption and environmental effort. The model is employed to investigate the determinants of the turning point and the cost effectiveness of different public policies aimed at a reduction of the environmental burden. Moreover, the model offers a potential explanation of an N-shaped pollution–income relation. It is shown that the model is compatible with most empirical regularities on economic growth and the environment.

Keywords: abatement, economic growth, environmental Kuznets curve, external effects, pollution, public policy

1. Introduction

The Environmental Kuznets Curve (EKC) hypothesis states that there is an inverted U-shaped relationship between environmental degradation and the level of income. Starting with Grossman and Krueger (1993) this pattern has been intensively debated in empirical terms; recent reviews are provided by Dasgupta et al. (2002) and Stern (2004). The EKC has also captured considerable attention from policymakers and theorists. This is due to the fact that the EKC hypothesis implies that pollution diminishes once a critical threshold level of income is reached. As a consequence, there is the hope that – loosely speaking – the environmental problem sooner or later peters out as the economy grows.

There are two major strands within the theoretical EKC literature. In the first class of models an EKC arises from shifts in the use of production technologies, which differ in their pollution intensity (Stokey 1998; Smulders et al. 2005). The second class focuses on the characteristics of the abatement technology (John and Pecchenino 1994; Selden and Song 1995; Andreoni and Levinson 2001; Chimeli and Braden 2002; Brock and Taylor 2004).

In an important paper, Andreoni and Levinson (2001) (thereafter AL) set up a static model to show that an EKC can be explained with increasing returns to scale (IRS) in the abatement technology. This approach can be viewed as a reduced

L. Bretschger and S. Smulders (eds.), Sustainable Resource Use and Economic Dynamics, 17–34.

form of a large number of models which focus on very different mechanisms (e.g. a shift in technology or a shift in institutions).

The level of income at which pollution peaks (labelled "the turning point") and the associated level of pollution are of fundamental interest from the perspective of public policy. A sound understanding of the pollution–income relation (PIR) could provide important information for public policies aimed at a reduction of the environmental burden. The empirical EKC literature has accordingly devoted much effort to the determination of this critical threshold. The results show, however, a large dispersion across different studies. For instance, the reported turning points for sulphur dioxide range from \$2,900 to \$908,200 and for nitrogen oxides from \$5,500 to \$30,800 (in 1985 PPP\$; Lieb 2003). Given these diverse empirical results, it is clearly desirable to better understand the determinants of the turning point from a theoretical perspective.

We set up a simple dynamic EKC model with the following characteristics: Pollution is a by-product of consumption activities, it is modelled as flow pollution and it creates disutility. Households can spend resources on abatement to reduce gross pollution. Following AL we assume that there are IRS in abatement giving rise to an EKC. There are two market distortions due to external effects associated with consumption and abatement activities. Permanent growth results from an accumulable stock of capital and a linear final-output technology.

The paper at hand focuses on two issues: First, we employ the simple dynamic EKC model to better understand the determinants of the turning point. The factors which are of major interest in this type of models are the preference for a clean environment, the degree of IRS in abatement and the magnitude of external effects. Second, we investigate the effectiveness of public policy measures aimed at a reduction of the environmental burden. In this context, it is important to have a model with multiple market failures so that the question of relative policy effectiveness can be studied.

Pollution is modelled as flow pollution. The reason lies in the fact that an EKC is more likely to arise for flow pollutants than for stock pollutants. This is best illustrated by Lieb (2004, p. 484) who reports that "*almost all studies agree that there is an EKC for sulphur dioxide* (SO_2), *suspended particulate matter (SPM), oxides of nitrogen* (NO_x), *carbon monoxide (CO), and for some (but not all) sorts of river pollution (PR) ... Although all these pollutants are stock pollutants, they all have short life-times and can therefore be considered as flow pollutants from a long-run point of view.*"

There are a number of theoretical papers on the EKC which consider the determinants of the turning point; some of these papers also investigate the role of public policies. Brock and Taylor (2004) use an augmented Solow model to demonstrate that an EKC arises along the transition to the steady state. Although there is polluting production in this model, there is no market failure. Lieb (2004) uses an overlapping generations model with a stock pollution and a flow pollution. He focuses on the different pollution paths of the stock and the flow pollution.

The model captures several external effects associated with production and abatement. However, only the problem of a myopic government is analysed implying that the intragenerational externalities are internalised, while the intergenerational externalities are not. Moreover, the effectiveness of public policy measures is not considered since the unregulated market economy is not investigated. Chimeli and Braden (2002) employ a simple endogenous growth model with environmental quality. They show that environmental quality follows a V-shaped pattern, thereby explaining an EKC for a stock pollution. There is single external effect associated with polluting production. Hence, the consequences of multiple external effects cannot be studied. Finally, Anderson and Cavendish (2001) employ a dynamic simulation model to investigate the consequences of public policy measures on the turning point. This computable equilibrium model has the advantage of being able to directly include different aspects of the real world which are important in this context. However, general equilibrium feedback effects are excluded and optimal taxes cannot be derived.

In section 2, the basic AL model is sketched. In section 3, a simple dynamic EKC model is set up. The decentralised and the centralised solution are investigated and the optimal tax scheme is determined. In section 4, a parameterised version of the model is employed to investigate the determinants of the turning point and the relative effectiveness of public policies. In section 5, it is shown that the model can potentially explain an N-shaped PIR. Section 6 demonstrates that the model is compatible with important stylised facts on economic growth and the environment. Section 7 summarises and concludes.

2. The Andreoni and Levinson EKC Model

The AL (2001) model is sketched to provide a reference point for the following discussion. Utility of the representative agent depends positively on consumption C and negatively on pollution P. The utility function is:

$$U = U(C, P). \tag{1}$$

Pollution is a function of C and environmental effort E according to:

$$P = C - B(C, E). \tag{2}$$

Pollution increases one-to-one with consumption (gross pollution), the first term on the RHS. On the other hand, pollution decreases due to abatement, the second term. The abatement technology $B(C, E)$ is increasing in both arguments. Both "inputs" are essential for abatement, i.e. $B(0, E) = B(C, 0) = 0$. Finally, the resource constraint is $Y = C + E$, where Y denotes available resources.

There are two conditions which together guarantee the existence of an EKC (AL 2001, p. 277). The first – related to preferences – states that "*the marginal willingness to pay to clean up the last speck of pollution does not go to zero as income approaches infinity*". This rather weak condition is easily satisfied since pollution abatement can be regarded as a normal good.[1] The second condition – related to abatement technology – states that there must be IRS in abatement.

Using $U(C, P) = C - zP$ with $z = 1$ and $B(C, E) = C^{\alpha} E^{\beta}$, AL show that an EKC results provided that $\alpha + \beta > 1$. This can be seen by inspecting the pollution function in terms of Y:

$$P(Y) = \frac{\alpha}{\alpha + \beta} Y - \left(\frac{\alpha}{\alpha + \beta}\right)^{\alpha} \left(\frac{\beta}{\alpha + \beta}\right)^{\beta} Y^{\alpha+\beta}. \tag{3}$$

The preceding equation results from $P = C - C^{\alpha} E^{\beta}$, $C^* = \frac{\alpha}{\alpha+\beta} Y$ and $E^* = \frac{\beta}{\alpha+\beta} Y$, where C^* and E^* are the optimal levels of C and E. Equation (3) implies that $P(Y)$ is concave in Y provided that $\alpha + \beta > 1$. Hence, IRS in abatement ($\alpha + \beta > 1$) represent a necessary condition for the existence of an EKC.

3. A General Dynamic EKC Model

A simple dynamic EKC model is set up. Pollution results as a by-product of consumption activities and is modelled as flow pollution. Households can reduce pollution by spending resources on abatement. The abatement technology is characterised by IRS, which gives rise to an EKC. There is a homogeneous final-output good which is produced under constant returns to scale using (physical and human) capital as the sole input factor. Households earn income by renting capital to firms. Output and factor markets are perfectly competitive. We consider two types of externalities and hence the decentralised solution diverges from the centralised solution. At first, the market economy is considered and then the centralised solution is investigated. Finally, the optimal tax scheme is determined.[2]

3.1. THE DECENTRALISED ECONOMY

There is a large number of identical households ordered on the interval [0,1]. The representative household derives utility from consumption C and disutility from net pollution P. The instantaneous utility function is $U(C, P)$ with $U_C > 0$, $U_{CC} < 0$, $U_P < 0$ and $U_{PP} < 0$.[3] The flow of pollution (per period of time) is the difference between gross pollution $G(C, \bar{C})$ and abatement $B(C, E, \bar{E})$:

$$P(C, \bar{C}, E, \bar{E}) = G(C, \bar{C}) - \min\{B(C, E, \bar{E}), G(C, \bar{C})\}, \tag{4}$$

where E is environmental effort and a “bar” above a variable denotes its economywide average. Pollution is modelled to result from consumption.[4] Note that this definition implies that pollution cannot turn negative, which is appropriate for a pure flow pollution (Lieb 2004, p. 488; Egli 2005).

Direct examples for polluting consumption activities would be the use of automobiles and central heating. Turning to environmental effort, we can interpret the model in the sense that both households as well as firms conduct abatement. It is, however, plausible and convenient to let the incidence of abatement costs fall on households. To clarify this aspect, consider a real-world example: Abatement in the case of driving automobiles comprises the installation of catalytic converters and strainers. Although the major part of this abatement activity (development and installation) is conducted by firms, households face the decision for, and bear the costs of this environmental effort.

There are two kinds of externalities: First, polluting consumption is only partially taken into account by the representative household, i.e. there is a (negative) pollution externality. Second, environmental effort aimed at reducing (net) pollution affects also the society as a whole, i.e. there is a (positive) externality resulting from environmental effort. As an example, consider again the use of automobiles. It is the household who bears the financial burden but it is society that primarily benefits from the implementation of catalytic converters and strainers. External effects are associated with $\bar{C}$ and $\bar{E}$.

Let r denote the rental price of capital K owned by the representative household, who earns income of rK. Gross expenditures (including taxes) are given by $(1+\tau_C)C + (1+\tau_E)E$, where τ_C and τ_E represent taxes (subsidies) on C and E. Tax revenues T are redistributed in a lump-sum manner according to a balanced-budget rule, i.e. $T = \tau_C C + \tau_E E$. Households maximise the present value of an infinite utility stream. The associated dynamic problem may be expressed as (time index suppressed):

$$\max_{\{C,E\}} \int_0^\infty U(C,P)e^{-\rho t}dt \tag{5}$$

$$s.t. \quad P(C,\bar{C},E,\bar{E}) = G(C,\bar{C}) - \min\{B(C,E,\bar{E}), G(C,\bar{C})\} \tag{6}$$

$$\dot{K} = rK - (1+\tau_C)C - (1+\tau_E)E + T \tag{7}$$

$$K(0) = K_0, \tag{8}$$

where ρ is the time preference rate, t the time index, $\dot{K}$ the rate of change of K per period of time and K_0 the initial stock of capital, respectively.

The (current-value) Hamiltonian for this problem reads:

$$H = U[C, P(C,\bar{C},E,\bar{E})] + \lambda[rK - (1+\tau_C)C - (1+\tau_E)E + T], \tag{9}$$

where λ is the shadow price of K. The necessary conditions are given by:[5]

$$\frac{U_C + U_P P_C}{1+\tau_C} = \lambda \tag{10}$$

$$\frac{U_P P_E}{1+\tau_E} = \lambda \tag{11}$$

$$\dot{\lambda} = -\lambda(r-\rho), \tag{12}$$

where U_x and P_x denote the partial derivatives of U and P with respect to $x \in \{C, E\}$, respectively. For ease of interpretation, assume that $\tau_C = \tau_E = 0$. Equation (10) then shows that along the optimal growth path the (private) marginal utility of consumption must equal λ. Marginal utility of consumption comprises two components: (i) direct utility from consumption U_C and (ii) disutility from pollution $U_P P_C$. Remember that P_C captures a gross pollution effect G_C and an abatement effect B_C. Similarly, (11) indicates that marginal utility from environmental effort $U_P P_E$ must equal λ. Equation (12) shows that λ vanishes at rate $r - \rho > 0$.

The representative final-output firm produces a homogeneous good using capital only. The constant returns to scale technology is $Y = AK$, where Y is final output and A a constant technology parameter. Capital depreciates at constant rate $\delta \geqslant 0$. Output maximisation implies $r = A - \delta$.

3.2. THE CENTRALISED ECONOMY

The social planner maximises welfare of the representative individual, taking the external effects into account. The associated problem reads:

$$\max_{\{C,\bar{C},E,\bar{E}\}} \int_0^\infty U(C, P)e^{-\rho t} dt \tag{13}$$

$$s.t. \quad P(C, \bar{C}, E, \bar{E}) = G(C, \bar{C}) - \min\{B(C, E, \bar{E}), G(C, \bar{C})\} \tag{14}$$

$$\dot{K} = F(K) - \delta K - C - E \tag{15}$$

$$K(0) = K_0. \tag{16}$$

The (current-value) Hamiltonian reads:

$$H = U[C, P(C, \bar{C}, E, \bar{E})] + \lambda[F(K) - \delta K - C - E] \tag{17}$$

and the necessary conditions are given by:[6]

$$U_C + U_P(P_C + P_{\bar{C}}) = \lambda \tag{18}$$

$$U_P(P_E + P_{\bar{E}}) = \lambda \tag{19}$$

$$\dot{\lambda} = -\lambda(F_K - \delta - \rho). \tag{20}$$

Comparing (18) and (19) to (10) and (11) shows the differences between the two solutions. When deciding on the optimal levels of C and E the social planner, in contrast to the private agent, takes the external consequences associated with $\bar{C}$ and $\bar{E}$ into account. Specifically, the social planner considers also the effects of $\bar{C}$ on gross pollution ($U_P P_{\bar{C}} = -U_P G_{\bar{C}}$) as well as the consequences of $\bar{E}$ on abatement ($U_P P_{\bar{E}} = -U_P B_{\bar{E}}$).

3.3. OPTIMAL TAX SCHEME

Comparing (18) and (19) to (10) and (11) yields the optimal tax scheme:

$$\tau_C^* = -\frac{U_P P_{\bar{C}}}{U_C + U_P(P_C + P_{\bar{C}})} > 0 \tag{21}$$

$$\tau_E^* = -\frac{P_{\bar{E}}}{P_E + P_{\bar{E}}} < 0. \tag{22}$$

Equation (22) shows that the optimal subsidy on environmental effort τ_E^* equals the ratio of the external marginal effect of environmental effort on pollution $P_{\bar{E}} < 0$ and the overall (private and external) marginal effect of environmental effort on pollution $P_E + P_{\bar{E}} < 0$. Similarly, the optimal consumption tax τ_C^* is the ratio of the external marginal consumption effect on utility $U_P P_{\bar{C}} < 0$ and the overall marginal consumption effect on utility given by $U_C + U_P(P_C + P_{\bar{C}}) > 0$.[7] Consider the consequences of a consumption tax on the decisions of the representative household. Implementing $\tau_C > 0$ reduces the LHS of (10). Holding λ constant, (10) then requires the marginal utility of consumption to increase. This calls for a reduction of C. An analogous interpretation (with $\tau_E < 0$) applies to (11).

4. A Specific Dynamic EKC Model

A parameterised version of the model is employed to investigate the determinants of the turning point and the effectiveness of public policies. At first, we consider the centralised solution. Subsequently, we turn to the more relevant case of an unregulated/imperfectly regulated economy.

4.1. PARAMETERISATION

We parameterise instantaneous utility $U(C, P)$, gross pollution $G(C, \bar{C})$ and abatement $B(C, E, \bar{E})$ as follows:

$$U(C, P) = \log(C - zP) \text{ with } z > 0, C \geqslant zP \quad (23)$$

$$G(C, \bar{C}) = C^{1-\omega}\bar{C}^{\omega} \text{ with } 0 < \omega < 1 \quad (24)$$

$$B(C, E, \bar{E}) = C^{\alpha}E^{\beta}\bar{E}^{\eta} \text{ with } 0 < \alpha, \beta, \eta < 1, \quad (25)$$

where z shows the desire for a clean environment. A lower value of z means that a given amount of pollution causes less disutility and individuals will accordingly spend more on C and less on E. In (24) $C^{1-\omega}$ represents the internal effect of consumption on gross pollution and $\bar{C}^{\omega}$ is the corresponding external effect. Similarly, E^{β} is the private and $\bar{E}^{\eta}$ the external effect of environmental effort in abatement.[8]

A short explanation of the instantaneous utility function (23) is indicated. Since $C = \bar{C}$ and $E = \bar{E}$, pollution is $P = C - C^{\alpha}E^{\beta+\eta}$. Moreover, assuming $z = 1$ one gets $U[C, P(C, E)] = \log(C^{\alpha}E^{\beta+\eta})$.[9] This formulation has the advantage that C and E enter utility additively separable, which enables an analytical solution for the social planner's problem.

4.2. ANALYTICAL RESULTS

The PIR is derived analytically and the determinants of the turning point are discussed. The focus is on the centralised solution with $z = 1$, which allows derivation of analytical results.

4.2.1. *The time path of pollution $P(t)$ and the* PIR $P(Y)$

The model under study is an augmented AK-model which implies:

$$K = K_0 e^{(A-\delta-\rho)t} \tag{26}$$

$$\lambda = \frac{\alpha+\beta+\eta}{K_0\rho} e^{-(A-\delta-\rho)t}. \tag{27}$$

From (18), (19), (27) and (23) to (25) one gets:

$$P(t) = \frac{K_0 e^{(A-\delta-\rho)t}\alpha\rho}{\alpha+\beta+\eta} - \left(\frac{K_0 e^{(A-\delta-\rho)t}\alpha\rho}{\alpha+\beta+\eta}\right)^{\alpha} \left(\frac{K_0 e^{(A-\delta-\rho)t}(\beta+\eta)\rho}{\alpha+\beta+\eta}\right)^{\beta+\eta} \tag{28}$$

Furthermore, the PIR may be expressed as follows:

$$P(Y) = cY - (cY)^{\alpha}(hY)^{\beta+\eta}, \tag{29}$$

where $c := \frac{C}{Y}$ is the consumption rate and $h := \frac{E}{Y}$ the "environmental effort rate". From $\hat{K} = A - \delta - \rho = A - \delta - C/K - E/K$ and the parameterised versions of (18) and (19) one gets:

$$c = \frac{\alpha\rho}{A(\alpha+\beta+\eta)} \quad \text{and} \quad h = \frac{(\beta+\eta)\rho}{A(\alpha+\beta+\eta)}. \tag{30}$$

The PIR is illustrated in Figure 1(a) and the time path of pollution in Figure 1(b). These graphs use the baseline set of parameters (section 3). As in AL (2001), IRS in abatement is a necessary condition for a hump-shaped PIR.[10]

Figure 1(a) shows that pollution first rises with income, then declines and eventually becomes zero. This EKC represents a balanced growth phenomenon. Although pollution does not grow at constant rate, the illustrated pollution path represents a balanced growth phenomenon since pollution results from two endogenous variables (C and E), which both grow at constant rates. The required time span until pollution reaches its peak and becomes zero is quite long. The "EKC story" takes nearly 250 years, as displayed in Figure 1(b).

The EKC pattern displayed in Figure 1(a) is in line with empirical evidence reported by Grossman and Krueger (1995), which indicates that the PIR is asymmetric with an upper tail that declines relatively gradually.

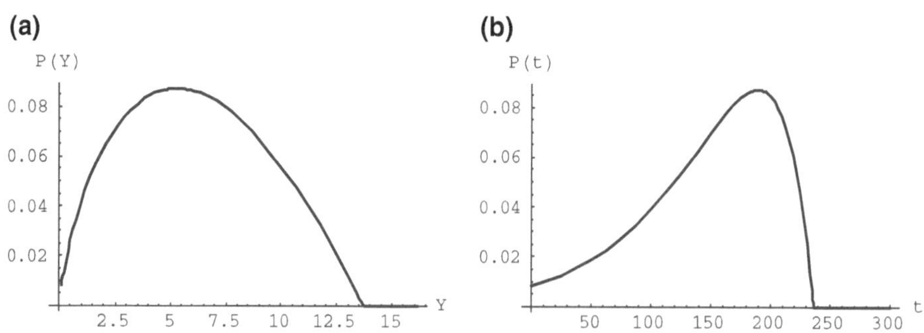

Figure 1. $P(Y)$ and $P(t)$ with IRS in abatement $(\alpha+\beta+\eta > 1)$.

4.2.2. *The Turning Point*

The determinants of the turning point are analysed. Closed-form solutions can only be obtained for the centralised economy with $z = 1$. Under these restrictions one can investigate the impact of basic technology and preference parameters on the turning point analytically. This represents an interesting limiting case, which is relevant in the sense that the qualitative results largely hold true also for the decentralised economy with $z < 1$, as investigated in section 4.3.

The point in time at which pollution reaches its maximum (t^*) is:

$$t^* = -\frac{\log[K_0^{\alpha+\beta+\eta-1}\alpha^{\alpha-1}(\beta+\eta)^{\beta+\eta}(\alpha+\beta+\eta)^{2-\alpha-\beta-\eta}\rho^{\alpha+\beta+\eta-1}]}{(\alpha+\beta+\eta-1)(A-\delta-\rho)}. \tag{31}$$

Note that z and ω do not appear on the RHS due to the restriction $z = 1$ and the fact that $C = \bar{C}$. The turning point in terms of income (Y^*) is:[11]

$$Y^* = \frac{A\alpha^{\frac{1-\alpha}{\alpha+\beta+\eta-1}}(\beta+\eta)^{-\frac{\beta+\eta}{\alpha+\beta+\eta-1}}(\alpha+\beta+\eta)^{1-\frac{1}{\alpha+\beta+\eta-1}}}{\rho}. \tag{32}$$

This critical income level is determined by the marginal product of capital A, the rate of time preference ρ, the elasticity of consumption in abatement α as well as the elasticities of environmental effort in abatement β and η. It is independent of δ and K_0.

Table I shows the comparative static results.[12] The first row shows that Y^* increases with A. For ease of interpretation, let us assume that $\alpha = \beta + \eta$ such that $C = E$.[13] In this case, the level of pollution depends only on consumption. Since an increase in A reduces the consumption rate [see (30)], the required level of income for pollution to reach its maximum increases. The second row indicates that Y^* falls as ρ rises. An analogous reasoning is applicable here. The rate of consumption rises with ρ [see (30)] and hence the required level of income for pollution to reach its maximum falls. The signs of the partial derivatives of Y^* with respect to α and β are indetermined.[14] In most instances, the derivatives with respect to α and β are negative. An increase in the degree of IRS in abatement leads, ceteris paribus, to a higher abatement output for each level of income and hence to a lower turning point. However, a positive sign cannot be excluded in

Table I. Comparative static results for Y^*

	$\frac{\partial Y^*}{\partial x}$ for $x \in \{A, \rho, \alpha, \beta\}$	
A	$Y^* \frac{1}{A}$	> 0
ρ	$Y^* \frac{-1}{\rho}$	< 0
α	$Y^* \frac{(\gamma-1)(-\alpha+\beta+\eta)+\alpha\gamma(\log[\gamma]+(\beta+\eta)(\log[\beta+\eta]-\log[\alpha])}{\alpha\gamma(\gamma-1)^2}$	?
β	$Y^* \frac{2+\gamma(\log[\gamma]-2)+\gamma(\alpha-1)(\log[\alpha]-\log[\beta+\eta])}{\gamma(\gamma-1)^2}$	?

Table II. Baseline set of parameters

Final output technology	$A = 0.12; \delta = 0.06$
Preferences	$\rho = 0.04$
Abatement technology	$\alpha = 0.6; \beta = 0.45; \eta = 0.05$
Gross pollution	$\omega = 0.1$

general; for instance, under the restrictions $\alpha = \beta + \eta$ and $z = 1$ the derivative with respect to α is positive.[15]

4.3. NUMERICAL ANALYSIS

The preceding analysis focused on the centralised solution with $z = 1$ implying that consumption and pollution have the same weight in the utility function. We now investigate the importance of external effects, the effectiveness of public policies and the implications of different environmental preferences, allowing for $z < 1$. To accomplish this task, the transition process is simulated using the backward integration procedure (e.g. Brunner and Strulik 2002).

4.3.1. *Calibration*

Table II shows the employed baseline set of parameters. The time preference rate ρ and the depreciation rate δ are similar to the parameter values used in previous exercises (e.g. Ortigueira and Santos 1997; Eicher and Turnovsky 2001). Given these values A is chosen such that the implied net rate of return on capital $(A - \delta)$ and the growth rate of per capita income $(A - \delta - \rho)$ are in line with empirically plausible numbers (6% and 2%). We choose ω and η such that the relative external effect of consumption in pollution (ω) and the relative external effect of environmental effort in abatement $\left(\frac{\eta}{\beta+\eta}\right)$ are both 10%, implying fairly moderate external effects.

We assume $\alpha + \beta + \eta > 1$. As in AL (2001), IRS in abatement are necessary for an EKC. This is in line with Xepapadeas (1994), where IRS in the pollution abatement sector (due to knowledge spillovers) is a necessary condition for unbounded growth without excess pollution (similarly Michel 1993). Alternatively, IRS in abatement may result from technological progress in the abatement technology (Anderson and Cavendish 2001). Regarding the empirical evidence, AL (2001, p. 281) argue that "*at the level of US states, average pollution abatement costs per dollar of GSP [gross state product] decline with industry size, across states and industries, and over time.*" Maradan and Vassiliev (2005) report that the marginal opportunity costs of carbon dioxide abatement are negatively associated with income. Moreover, β and η crucially determine the ratio of abatement expenditures and income, which ranges from about 3% for $z = 0.5$ to 15% for $z = 1$. These values are in line with the empirical figures reported by Brock and Taylor (2004, p. 6).

Table III. Elasticities of Y^* with respect to model parameters; unregulated economy ($\theta = 0$)

	ω	η	A	ρ	α	β	z
$Y^* z = 1$	0.67	−0.79	0.97	−0.90	−4.41	−5.74	−4.70
$Y^* z = 0.75$	0.46	−1.45	0.98	−0.90	−7.48	−7.40	−4.42
$Y^* z = 0.5$	0.28	−2.22	0.99	−0.91	−9.06	−8.61	−4.19

Table IV. Elasticities of Y^* with respect to model parameters; imperfectly regulated economy ($\theta = 0.5$)

	ω	η	A	ρ	α	β	z
$Y^* z = 1$	0.30	−0.75	0.99	−0.90	−2.71	−4.87	−4.98
$Y^* z = 0.75$	0.21	−1.46	1.00	−0.91	−6.92	−7.00	−4.60
$Y^* z = 0.5$	0.14	−2.27	1.00	−0.91	−8.90	−8.43	−4.29

4.3.2. *The turning point*

The dependence of Y^* on the model parameters is investigated numerically. Three different values of z are considered. In addition, the unregulated economy (Table III) is distinguished from an imperfectly regulated economy (Table IV).[16] We focus on these two cases since we believe that the real world is best represented by an unregulated or imperfectly regulated economy. The basic assumption here is that politicians know the optimal taxes but due to imperfections in the political process do not fully implement optimal taxes. The numbers reported in Tables III and IV show the elasticities of Y^* with respect to different model parameters, i.e. $\frac{\Delta Y^*/Y^*}{\Delta x/x}$ with $x \in \{\omega, \eta, A, \rho, \alpha, \beta, z\}$.[17]

Three points should be noted: First, the case of $z = 1$ is qualitatively identical to the cases of $z < 1$. By lowering z, the results change only gradually. The respective elasticities show the same sign for the unregulated economy (Table III) and for the imperfectly regulated economy (Table IV). Second, the analytical results from Table I are confirmed and the ambiguous effects of α and β are determined, at least numerically. Third, compared to the case investigated above (centralised solution with $z = 1$) the impact of additional model parameters (ω and η) can be assessed.

The first column of Table III shows the elasticity of Y^* with respect to ω. The displayed positive impact can be explained as follows: Since the gross pollution function is linear, the level of centralised C remains constant. Increasing ω leads to a larger gap between the centralised and the decentralised allocation. This implies that decentralised C rises, which, ceteris paribus, causes a higher level of P at each level of income. Graphically speaking, the EKC is expanded outwards and the turning point increases. This column also shows that the impact of ω on Y^* increases with z. A higher value of z (i.e. greener preferences) leads to a larger gap between the centralised and the decentralised solution, as can be seen by inspecting (18). This implies that the strength of the mechanism described above is reinforced. Finally, the effect of ω on Y^* is smaller for the imperfectly regulated economy (Table IV).

The second column of Table III gives the impact of a variation in η on Y^*, which is negative. An increase in η has two separate effects: First, environmental effort falls. To understand this effect, consider the case of a variation in η assuming that $\beta + \eta =$ constant. This implies that centralised E remains constant. Since the magnitude of the distortion increases, the gap between the centralised and the decentralised solution gets larger. Hence, E must decrease implying that pollution rises at each level of income and that the turning point increases as well. Second, by holding β fixed (assumed in Tables III and IV), an increase in η leads to a higher degree of IRS causing pollution to fall at each level of income. Consequently, the turning point decreases. The second effect dominates the first and hence the sign of this elasticity is negative.[18]

The third column (A) and the fourth column (ρ) are in line with the analytical results obtained from the special case investigated in section 4.2. The fifth (α) and sixth column (β) contain negative values. Increasing either α or β increases the degree of IRS in abatement, which has a strong negative impact on the turning point.[19] Finally, the last column (z) shows that an increase in z has a strong negative impact on Y^*.

4.3.3. *The cost effectiveness of public policies*

So far we have considered first-best policies in an unconstrained welfare-maximising setting. We now turn to a cost-effectiveness analysis. It is argued that there is a maximum level of pollution, which should not be exceeded. This threshold is determined outside the economic model under study and might be the result of ecological considerations; it is not determined by cost-benefit analysis. Moreover, to simplify matters, it is assumed that the regulator has only one policy instrument available in order to cap pollution. Specifically, this objective can be achieved by either implementing a tax on polluting consumption or a subsidy on environmental effort. This analysis aims to shed light on the following question: Is it optimal to primarily avoid pollution by taxing consumption, or is it instead optimal to primarily correct the problem of pollution by subsidising abatement activities?[20]

As a first step in trying to answer this question, we conduct the following policy experiment. The social planner implements the second-best policy (either τ_C or τ_E) taking the optimal behaviour of the private sector into account such that the constraint $P \leqslant P_{\max}$ holds. Implementing a second-best consumption tax while setting the subsidy on environmental effort equal to zero is labelled a τ_C-regime. The reverse situation is labelled a τ_E-regime. Figure 2 illustrates the result of this policy experiment based on the baseline set of parameters (Table II) and assuming that $K_0 = 70$, $z = 0.8$ and $P_{\max} = 0.6$. Figure 2(a) shows the resulting EKC under both a τ_C-regime and a τ_E-regime. Figure 2(b) shows the respective pollution paths along the time dimension. Note that the dashed segments of the respective EKCs are not realised. At those points in time where $P \leqslant P_{\max}$ becomes

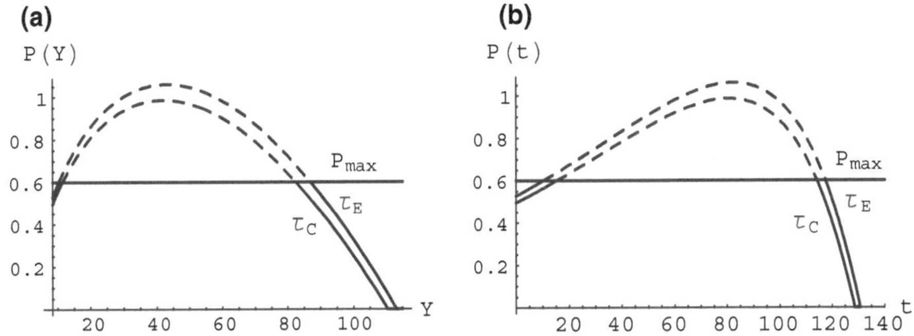

Figure 2. Comparative policy analysis.

binding, the policy instruments are adjusted such that pollution remains below P_{max}.

Compared to the τ_C-regime the τ_E-regime leads to a welfare gain which is equivalent to a permanent increase in consumption of 3.6%. This is a non-negligible number. To understand this result two points should be noted: First, Figure 2(b) shows that pollution is higher at each point in time under the τ_E-regime, which affects welfare negatively. Second, however, the τ_E-regime leads to a higher level of consumption than the τ_C-regime (not shown), which is reasonable since the τ_C-regime aims at discouraging polluting consumption. For the underlying set of parameters the second effect dominates. It is clear that this result is especially sensitive with respect to the parameter capturing the preference for a clean environment. Since the τ_C-regime leads to both lower P and C, whereas the τ_E-regime is associated with higher P and C, an increase in z reduces or may even reverse the advantageousness of the τ_E-regime. For instance, for $z = 0.82$ the advantage of the τ_E-regime vis-à-vis the τ_C-regime is reduced to a welfare gain equivalent to a permanent increase in consumption of 1.7%.[21]

5. N-shaped Pollution–income Relation

A number of empirical studies argue that the PIR is N-shaped, at least for some pollutants (Grossman and Krueger 1995, section IV; Lieb 2003). This is important because, in this case, pollution eventually increases with income.

The model under study provides a potential explanation for this phenomenon. Imagine the economy develops at first along the upward sloping branch of the EKC resulting from the market economy (see Figure 3). At some point in time, policy instruments are implemented and pollution diminishes. In the model, the economy jumps to the centralised EKC; in reality this process is distributed over time. Provided that the economy is still below Y^* of the centralised solution, pollution starts to increase again. This produces an N-shaped PIR resulting from the interplay of public policy and the intrinsic properties of the model. Note that this explanation implies in fact an M-shaped PIR. As soon as the peak of pollution (on the centralised EKC) is reached, pollution starts to decline.

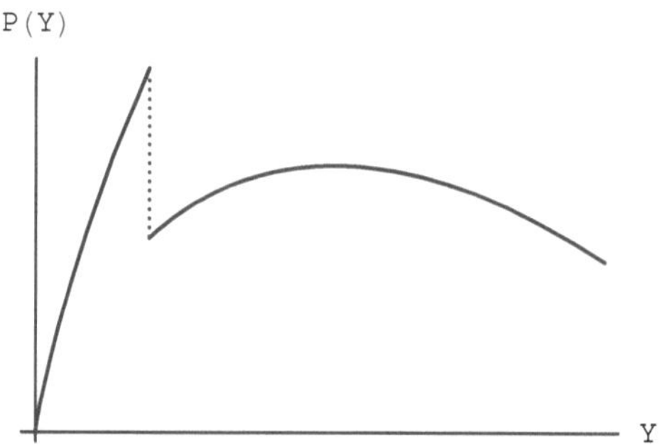

Figure 3. M-shaped PIR.

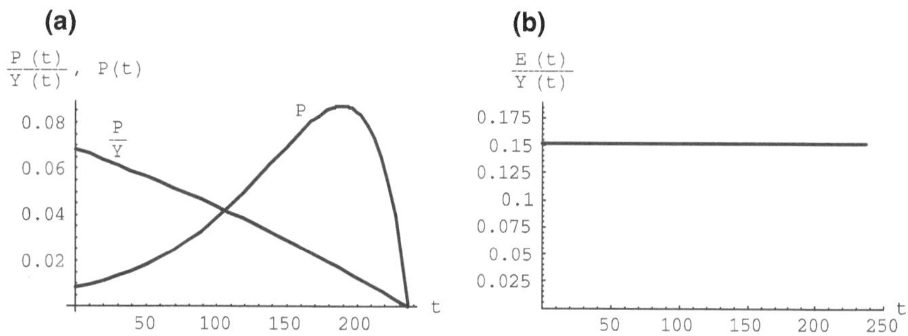

Figure 4. Comparison to empirical regularities.

Future empirical research aimed at explaining such a pattern should take this possibility into account. This explanation implies that the first downward movement is policy induced, i.e. it should succeed the implementation of environmental regulations aimed at a reduction of pollution. The subsequent increase in pollution is then simply due to the fact that growth might be accompanied by a rise in pollution. Moreover, an N-shaped pattern can result provided that there are less than IRS in abatement. Finally, Giles and Mosk (2003) find indeed an M-shaped EKC pattern using long-run data on methane emissions for New Zealand.

6. Other Empirical Regularities

A dynamic EKC model should not only reproduce an inverted U-shaped PIR. It should also be compatible with the remaining empirical regularities on economic growth and the environment. These have been reported by Brock and Taylor (2004) based on US data for 1950–2001: First, the emission intensities (P/Y in our notation) for most pollutants are declining over time. Second, despite the fact that emission intensities decline, the emission levels (P) continue to increase for a certain period of time. Third, abatement costs relative to GDP (E/Y) are roughly constant.

The above model is compatible with these empirical regularities. Figure 4(a) shows that the emission intensity (P/Y) is declining over time and that the pollution level (P) continues to increase for a certain period of time although pollution intensity is falling.[22] Figure 4(b) illustrates that abatement expenditures relative to GDP (E/Y) are constant over time.

The model, being an augmented AK growth model, is compatible with most of the Kaldor (1961) facts: the growth rate of per capita output, the capital-output ratio and the real rate of return on capital are constant.

7. Summary and Conclusions

We have set up a simple dynamic EKC model with multiple market failures resulting from external effects associated with polluting consumption and environmental effort. The model has been used to investigate the determinants of the level of income at which pollution starts to decline (turning point) as well as the relative effectiveness of public policy measures aimed at a reduction of the environmental burden. The main results can be summarised as follows:

(1) The turning point in the first-best solution is most strongly affected by the degree of IRS in abatement and the preference for a clean environment. In addition, in the decentralised economy, the magnitude of external effects associated with polluting consumption and environmental effort also has a substantial impact. This result indicates the importance of public policy measures for controlling pollution.

(2) For the case policy aims at a cap on pollution, which is determined by ecological factors without a cost-benefit analysis, we investigate the following question: Is it optimal to primarily avoid pollution by taxing consumption or is it instead optimal to primarily correct the problem of pollution by subsidising abatement activities? Provided that only one policy instrument is available, it turns out that a subsidy on environmental effort should be preferred vis-à-vis a tax on polluting consumption, unless the preference for a clean environment is relatively high.

(3) It has been shown that an N-shaped PIR, observable for some specific pollutants, can potentially be explained from the interaction of public policy measures and the intrinsic properties of the model. Although we do not consider this explanation to be valid in general, we think that this kind of reasoning should be taken into account in future empirical research aimed at explaining this pattern.

(4) In addition to the empirical EKC hypothesis, the dynamic EKC model under study is compatible with the remaining empirical regularities associated with economic growth and the environment (Brock and Taylor, 2004). Moreover, the model is also compatible with most of the stylised facts on economic growth due to Kaldor (1961).

Acknowledgements

We are grateful for helpful comments and suggestions from Geir Asheim, Lucas Bretschger, Karen Pittel, Sjak Smulders, Scott Taylor and three anonymous referees.

Notes

1. Lieb (2002) shows that the normality of environmental quality is a necessary condition for the existence of an EKC.
2. There are other general growth models with pollution and external effects (e.g. Smulders and Gradus, 1996).
3. We do not restrict the cross derivatives at this stage.
4. More frequently, pollution is modelled as a by-product of production (e.g. Xepapadeas 2006). There are, however, other theoretical studies, beside AL (2001), which assume that consumption generates pollution (e.g. John and Pecchenino 1994).
5. Since we are interested in an EKC, we consider "interior solutions" where $B < G$. In addition, the transversality condition $\lim_{t\to\infty} e^{-\rho t}\lambda K = 0$ must hold. We assume that the necessary conditions are also sufficient for a maximum of the utility functional.
6. Once again, the transversality condition $\lim_{t\to\infty} e^{-\rho t}\lambda K = 0$ must hold and we assume that the necessary conditions are also sufficient.
7. Notice that $U_C + U_P(P_C + P_{\bar{C}}) = \lambda > 0$.
8. An appendix, available upon request, shows that the parameterised Hamiltonian functions are concave such that the necessary conditions are also sufficient for a maximum of the utility functional.
9. The utility function requires $C - zP \geqslant 0$. For $z \leqslant 1$ this restriction is automatically satisfied since C is gross pollution and P is net pollution. Moreover, the utility function implies $U_{CP} = \frac{1}{(C-zP)^2} > 0$, which might appear counterintuitive. A rise in P has the same effect as a reduction in C, hence U_C increases with P. According to Michel and Rotillon (1995) $U_{CP} > 0$ can be interpreted as a compensation effect.
10. In a more general version of the AL (2001) model Plassmann and Khanna (2004, p. 16) show that "*for non-constant returns to scale in gross pollution, a sufficient condition for pollution to decline is rather that the returns to scale in abatement exceed the returns to scale in gross pollution.*"
11. This is basically the solution for Y^* one would obtain from the AL (2001) model.
12. To simplify notation, we define $\gamma = \alpha + \beta + \eta$.
13. A similar reasoning would apply to the case $\alpha \neq \beta + \eta$.
14. Since we are considering the centralised solution with $z = 1$, $\frac{\partial Y^*}{\partial \eta} = \frac{\partial Y^*}{\partial \beta}$.
15. In this case, the relevant range of consumption is $0 < C < 1$. Within this range an increase in α lowers, ceteris paribus, abatement output. As a result, the maximum level of pollution occurs at a higher C-level. With $\alpha = \beta + \eta$ the rate of consumption is independent of α and hence a higher C-level implies a higher Y^*.
16. The tax rates imposed are specified as $\tau_C = \theta_C \tau_C^*$ and $\tau_E = \theta_E \tau_E^*$, where $\tau_C^* > 0$ and $\tau_E^* < 0$ are optimal taxes (section 3.3); $\theta_C \geqslant 0$ and $\theta_E \geqslant 0$ indicate the extent of tax implementation.
17. The elasticities are based on an 10% increase of the parameter under consideration.
18. The results are nearly identical for the unregulated and the imperfectly regulated economy. This is due to the fact that the IRS argument does not depend on the degree of regulation.
19. As for the analytical solution the impact of δ is zero.
20. Ecologists usually argue in favour of the first strategy, whereas economists are more likely to prefer a combined strategy.
21. Although the welfare gain shrinks as z converges to unity, the τ_E-regime is preferable as long as $z < 1$.
22. Figure 4 is based on the centralised solution with $z = 1$ and the baseline set of parameters.

References

Anderson, D. and W. Cavendish (2001), 'Dynamic Simulation and Environmental Policy Analysis: Beyond Comparative Statics and the Environmental Kuznets Curve', *Oxford Economic Papers* **53**(4), 721–746.

Andreoni, J. and A. Levinson (2001), 'The Simple Analytics of the Environmental Kuznets Curve', *Journal of Public Economics* **80**(2), 269–286.

Brock, W.A. and M.S. Taylor, (2004), 'The Green Solow Model', *NBER Working Paper Series*, No. 10557.

Brunner, M. and H. Strulik (2002), 'Solution of Perfect Foresight Saddlepoint Problems: A Simple Method and Applications', *Journal of Economic Dynamics and Control* **26**(5), 737–753.

Chimeli, A. B. and J. B. Braden, (2002), 'The Environmental Kuznets Curve and Optimal Growth', *Working Paper*, Columbia University.

Dasgupta, S., B. Laplante, H. Wang and D. Wheeler (2002), 'Confronting the Environmental Kuznets Curve', *Journal of Economic Perspectives* **16**(1), 147–168.

Egli, H. (2005), 'A New Approach to Pollution Modelling in Models of the Environmental Kuznets Curve', *Swiss Journal of Economics and Statistics* **141**(3), 459–473.

Eicher, Th. S. and St. J. Turnovsky (2001), 'Transitional Dynamics in a Two Sector Non-Scale Growth Model', *Journal of Economic Dynamics and Control* **25**(1–2), 85–113.

Giles, D.E.A. and C. Mosk, (2003), 'Ruminant Eructation and a Long-Run Environmental Kuznets' Curve for Enteric Methane in New Zealand: Conventional and Fuzzy Regression Analysis', Department of Economics, University of Victoria, *Econometrics Working Paper*, EWP0306.

Grossman, G. M. and A. B. Krueger (1993), 'Environmental Impacts of a North American Free Trade Agreement', in P. M. Garber, ed., *The Mexico-U.S. Free Trade Agreement*. Cambridge and London: MIT Press.

Grossman, G. M. and A. B. Krueger (1995), 'Economic Growth and the Environment', *Quarterly Journal of Economics* **110**(2), 353–377.

John, A. A. and R. A. Pecchenino (1994), 'An Overlapping Generations Model of Growth and the Environment', *The Economic Journal* **104**(427), 1393–1410.

Kaldor, N. (1961), 'Capital Accumulation and Economic Growth', in F. A. Lutz and D. C. Hague, eds., *The Theory of Capital. Proceedings of a Conference held by the International Economic Association*. London: Macmillan.

Lieb, Ch. M. (2002), 'The Environmental Kuznets Curve and Satiation: A Simple Static Model', *Environment and Development Economics* **7**(3), 429–448.

Lieb, Ch. M., (2003), 'The Environmental Kuznets Curve. A Survey of the Empirical Evidence and of Possible Causes', University of Heidelberg, Department of Economics, *Discussion Paper Series*, No. 391.

Lieb, Ch. M. (2004), 'The Environmental Kuznets Curve and Flow versus Stock Pollution: The Neglect of Future Damages', *Environmental and Resource Economics* **29**(4), 483–506.

Maradan, D. and A. Vassiliev (2005), 'Marginal Costs of Carbon Dioxide Abatement: Empirical Evidence from Cross-Country Analysis', *Swiss Journal of Economics and Statistics* **141**(3), 377–410.

Michel, Ph., (1993), 'Pollution and Growth towards the Ecological Paradise', *Fondazione Eni Enrico Mattei, Nota di Lavoro 80.93*.

Michel, Ph. and G. Rotillon (1995), 'Disutility of Pollution and Endogenous Growth', *Environmental and Resource Economics* **6**(3), 279–300.

Ortigueira, S. and M. S. Santos (1997), 'On the Speed of Convergence in Endogenous Growth Models', *American Economic Review* **87**(3), 383–399.

Plassmann, F. and N. Khanna, (2004), 'A Note on "The Simple Analytics of the Environmental Kuznets Curve"', *Working Paper 0314*, Binghamton University, Department of Economics (forthcoming in *Environment and Development Economics)*.

Selden, Th. M. and D. Song (1995), 'Neoclassical Growth, the J Curve for Abatement, and the Inverted U Curve for Pollution', *Journal of Environmental Economics and Management* **29**(2), 162–168.

Smulders, S., L. Bretschger and H. Egli, (2005), 'Economic Growth and the Diffusion of Clean Technologies: Explaining Environmental Kuznets Curves', *Economics Working Paper Series 05/42*, ETH Zurich.

Smulders, S. and R. Gradus (1996), 'Pollution Abatement and Long-Term Growth', *European Journal of Political Economy* **12**(3), 505–532.

Stern, D. I. (2004), 'The Rise and Fall of the Environmental Kuznets Curve', *World Development* **32**(8), 1419–1439.

Stokey, N. L. (1998), 'Are there Limits to Growth?', *International Economic Review* **39**(1), 1–31.

Xepapadeas, A., (1994), 'Long-Run Growth, Environmental Pollution, and Increasing Returns', *Fondazione Eni Enrico Mattei, Nota di Lavoro 67.94*.

Xepapadeas, A., (2006), 'Economic Growth and the Environment', in K.-G. Mäler and J. Vincent, eds., *Handbook of Environmental Economics*, Vol. 3, Chapter 23, North-Holland: Elsevier Publishers.

3. The Optimal Timing of Adoption of a Green Technology

MARIA A. CUNHA-E-SÁ* AND ANA B. REIS
Faculdade de Economia, Universidade Nova de Lisboa, Campus de Campolide, P-1099-032, Lisbon, Portugal;
* *Author for correspondence (phone + 351-213801660; fax + 351-213870933; e-mail: mcunhasa@fe.unl.pt)*

Abstract. We study the optimal timing of adoption of a cleaner technology and its effects on the rate of growth of an economy in the context of an AK endogenous growth model. We show that the results depend upon the behavior of the marginal utility of environmental quality with respect to consumption. When it is increasing, we derive the capital level at the optimal timing of adoption. We show that this capital threshold is independent of the initial conditions on the stock of capital, implying that capital-poor countries tend to take longer to adopt. Also, country-specific characteristics, as the existence of high barriers to adoption, may lead to different capital thresholds for different countries. If the marginal utility of environmental quality decreases with consumption, a country should never delay adoption; the optimal policy is either to adopt immediately or, if adoption costs are "too high", to never adopt. The policy implications of these results are discussed in the context of the international debate surrounding the environmental political agenda.

Keywords: cost of adoption, growth, optimal timing of adoption, pollution, technology adoption

1. Introduction

Adoption of new technologies plays a very important role in economic growth, namely, by determining its pace and the rate of change of productivity, as discussed in Hall and Khan (2003). There is already a large literature studying the effect of the degradation of environmental quality, as an externality from production that decreases utility, on the optimal rate of growth.[1] When the negative effect of the decrease in environmental quality on welfare is taken into account, the optimal rate of growth always decreases, and it may even occur that optimal sustainable growth is not feasible, as in Stokey (1998). Considering endogenous growth models, it has been shown that (endogenous) technological change that allows an increase in production without decreasing environmental quality reduces its negative impact on the optimal rate of growth. Thus, we may say that the development of cleaner technologies can be important to reconciliate steady-state growth with care for the environment.

For many countries, in particular developing economies, the problem is mostly one of adopting technologies that have already been developed elsewhere. But even if the technology is already known, when a new technology is adopted the economy incurs a cost.[2] Developing countries in general do not use the cleaner technologies available in the developed world. Several factors make the diffusion

L. Bretschger and S. Smulders (eds.), Sustainable Resource Use and Economic Dynamics, 35–52.

of new technologies difficult, contributing to this outcome; costs of adoption are a crucial one. These may include the change of plant equipment, the acquisition of new machinery, the payment of property rights, besides a less efficient institutional framework that contributes to an increase in those costs.[3] In the case of this paper, we assume that the capital remaining after adoption will be operated with the new technology. Examples of these kinds of technological improvements are car pollution devices or scrubbers used in cement plants.

Developed countries have been trying to impose on developing ones more stringent environmental standards. A study by OECD (2001) shows that "industries had mainly imported cleaner technology because new domestic environmental standards forced them to do so", and the adoption of new environmental standards has in many cases been forced by developed economies in international agreements. For instance, in recent WTO renegotiation meetings the EU has been attempting to introduce environmental issues in the agenda, probably envisaging the possibility of using trade restrictions to impose their own environmental standards.[4] On the other hand, most developing countries have been resisting this attempt. The United Nations Secretary-General Kofi Annan has supported this position, arguing that environmental issues should not be used as pretexts for trade restrictions. He has pointed out that on the contrary "practical experience has shown that trade and investment not only bring economic development, but often bring higher standards of human rights and environmental protection as well" (UN, 1999). So, perhaps we should look for higher income in developing countries, as this would bring about environmental quality.

The optimal timing of adoption for developing countries may be different from the one developed economies try to impose on them. This raises the question of the legitimacy of developed countries attempting to impose their own environmental standards on poor countries, and may explain why these have been against those initiatives. This paper addresses this question by determining the optimal timing of adoption of a cleaner technology.

Our problem is related to the literature on optimal stopping (Dixit and Pindyck, 1994), in a context where the evolution of the state variable, the stock of capital, is not exogenous. We solve the problem by first studying the effects of an anticipated adoption, and then use the results obtained to determine the optimal timing of adoption, and the corresponding capital threshold. A similar approach was used in a different setup by Khan and Ravikumar (2002). These authors study the adoption of a more productive technology, in a context where capital accumulation has no externalities, in discrete time.

The results obtained depend upon the behavior of adoption benefits and costs over time. We show that the evolution of benefits over time is determined by the behavior of the marginal utility of environmental quality with respect to consumption. If it decreases, implying that the benefit from adoption decreases with the stock of capital, a country should never delay adoption: the optimal policy is either to adopt immediately or, if adoption costs are "too high", to never adopt. If the marginal utility of environmental quality increases with consumption, implying that the benefit from adoption increases with the stock of capital, we show that there is a threshold level of capital independent of the initial stock at which it is optimal to adopt. Consequently, all countries with the same characteristics will adopt at the same capital level, and capital-poor countries will take longer to adopt a cleaner technology. This suggests that capital flows from developed countries to developing ones may play an important role. These results are related to those derived in Brock and Taylor (2003) who study investment in abatement, and obtain that the capital threshold is independent of the initial conditions on the stock of capital, for the case of a flow pollutant. In the paper mentioned above, Khan and Ravikumar (2002), also find that delaying adoption may be optimal.

We also show that country-specific characteristics, as represented by the institutional framework or differences in the initial technological level, may give rise to different capital thresholds for different countries. Therefore, as developing countries have high barriers to adoption, and a less efficient institutional framework, they should be less eager to adopt. Hence, according to our results, developing countries are, in general, less willing to adopt cleaner technologies than developed ones.

Finally, we show that anticipated adoption of a cleaner technology decreases environmental quality in the period prior to adoption. This is related to the results in Reis (2001), where the case of a cleaner technology that may become available at no cost in the future is considered.

The remainder of the paper is organized as follows. Section 2 presents the model. Section 3 studies the adoption problem using a recursive approach. Section 4 derives the implications to the optimal behavior of the economy prior to adoption, and, finally, Section 5 concludes the paper. Technical details are presented in the Appendix.

2. The Model

We consider a closed economy with population normalized to one. The utility of the representative agent depends on per capita consumption C_t, and on the flow of environmental services Q_t, as follows:

$$U_t = \frac{(C_t Q_t^{\mu})^{1-\sigma} - 1}{1-\sigma}, 0 < \mu < 1, \sigma > 0 \tag{1}$$

For the utility function to be increasing and strictly concave in C and Q, the following restrictions are imposed: $\mu(1-\sigma) < 1$, and $\mu(1-\sigma) < \sigma$. The sign of the cross derivative depends on whether the inverse of the intertemporal elasticity of substitution, σ, is less than or greater than one.[5] When $\sigma > 1$, and the intertemporal elasticity of substitution is low, the marginal utility of environmental quality is a decreasing function of consumption, while for $\sigma < 1$, that is, when the intertemporal elasticity of substitution is high, the marginal utility of environmental quality is an increasing function of consumption.[6]

We consider the production function $Y = AK$ and zero depreciation of capital. Therefore,

$$\dot{K}_t = AK_t - C_t \tag{2}$$

Moreover, at each time period, the flow of environmental services is represented by

$$Q_t = a(AK_t)^{-\alpha}, \quad 0 < \alpha \leq 1 \tag{3}$$

where a is the quality of the technology.[7] Also, the higher the level of a, the higher is the quality embodied in Q_t for a given level of the stock of capital K_t, at each time period. Thus, an upgrading in a can be interpreted as an adoption of a greener technology by the country.

Our purpose is to study the optimal timing of adoption of a cleaner technology for this economy. So we consider the social planner's problem consisting of maximizing the present value of the stream of utility of the representative agent.

Taking into account that the timing of adoption, T, is also a decision variable of the policy maker, the infinite horizon problem for this economy can be stated as follows:[8]

$$\begin{aligned} V_0(K_0) = \max_{T,\{C_t\}} & \int_0^T \frac{(C_t Q_t^{\mu})^{1-\sigma} - 1}{1-\sigma} e^{-\rho t} dt \\ & + \int_T^{\infty} \frac{(C_t Q_t^{\mu})^{1-\sigma} - 1}{1-\sigma} e^{-\rho t} dt - (\gamma \upsilon + f) e^{-\rho T} \\ & \text{s.t.} \\ \dot{K}_t = {} & AK_t - C_t, \ \text{for } t \neq T \\ & K_0 \text{ given}, \ K_{T_+} = \beta K_{T_-}, \ \text{for } 0 < \beta \leq 1 \\ Q_t = {} & a_0(AK_t)^{-\alpha} \text{ for } t < T, \ \text{and } Q_t = a_1(AK_t)^{-\alpha} \text{ for } t > T \end{aligned} \tag{4}$$

where K_t is the stock of capital at time t, K_{T_+} represents the stock of capital immediately after adoption, and is defined by $K_{T_+} \equiv \lim_{t>T, t\to T} K_t$, while K_{T_-} is the stock of capital accumulated just before adoption takes place, that is, $K_{T_-} \equiv \lim_{t<T, t\to T} K_t$. Also, C_t is consumption at t, $a_1 = a_0(1 + \upsilon)$ represents the quality of technology after adoption, υ is the quality upgrade at T, and ρ is the discount rate. Moreover, $(\gamma\upsilon + f)$ represents the cost of adoption, where γ is the marginal cost of the jump, and f are fixed costs. Finally, due to adoption, part of the machinery may become obsolete. This translates into a decrease in the stock of capital at T, measured by $1 - \beta$.[9]

3. Technology Adoption: A Recursive Approach

The solution to (4) is obtained recursively. First, we solve for the period after adoption. Second, we solve for the period prior to adoption, taking as given the time of adoption. Finally, to determine the optimal T, we derive the value of adoption for each possible adoption date, T.

3.1. THE VALUE AFTER ADOPTION

When adoption has already occurred or in the case where it is not possible to adopt, and the quality level is never upgraded, the problem simplifies to

$$\begin{aligned} \varphi(K_T) &= \underset{\{C_t\}}{\text{Max}} \int_T^{\infty} \frac{(C_t Q_t^{\mu})^{1-\sigma} - 1}{1-\sigma} e^{-\rho(t-T)} dt \\ &\quad \text{s.t.} \\ \dot{K}_t &= AK_t - C_t \\ &\quad K_T \text{ given} \\ Q_t &= a_1 (AK_t)^{-\alpha} \end{aligned} \tag{5}$$

The optimal value function $\varphi(K_T)$ represents the value of the discounted stream of utility at $t = T$ obtained by operating with a given technology a_1.[10]

The corresponding current value Hamiltonian is as follows:

$$H = \frac{(C_t (a_1 (AK_t)^{-\alpha})^{\mu})^{1-\sigma} - 1}{1-\sigma} + \lambda_t [AK_t - C_t], \tag{6}$$

where λ_t represents the current shadow price of capital.

The first-order conditions for an interior solution are given by

$$a_1^{\mu(1-\sigma)} (AK_t)^{-\alpha\mu(1-\sigma)} C_t^{-\sigma} = \lambda_t \tag{7}$$

$$\dot{\lambda}_t = (\rho - A)\lambda_t + a_1^{\mu(1-\sigma)} (AK_t)^{-\alpha\mu(1-\sigma)} C_t^{-\sigma} \mu\alpha \frac{C_t}{K_t} \tag{8}$$

and the transversality condition is:

$$\lim_{t\to\infty} e^{-\rho t} \lambda_t K_t = 0 \tag{9}$$

The optimal solution for this problem implies that at each time t

$$\frac{C_t}{K_t} = \frac{\sigma - 1}{\sigma} A + \frac{\rho}{\sigma(1 - \alpha\mu)} \tag{10}$$

Therefore, there are no transitional dynamics and K and C grow at the same constant rate

$$\frac{\dot{C}_t}{C_t} = \frac{\dot{K}_t}{K_t} = \frac{A}{\sigma} - \frac{\rho}{\sigma(1 - \alpha\mu)}. \tag{11}$$

We assume that $(A(1-\alpha\mu)-\rho) > 0$, which is the condition for positive growth. If there were no pollution externalities, the optimal solution for this economy would imply a constant rate of growth equal to $(A-\rho)/\sigma$. As $\alpha\mu < 1$, the optimal rate of growth is lower in the presence of pollution, as in Bovenberg and Smulders (1995), Elbasha and Roe (1996), and Reis (2001), among others. From (10), the propensity to consume out of capital, C_t/K_t, is increasing in both μ and α. The higher the amount of pollution produced per unit of output, or the higher the concern for the environment represented by a larger μ, the lower is the growth rate of the economy, and, therefore, the greater the propensity to consume out of capital. In both cases, one additional unit of capital imposes a higher cost on utility, *ceteris paribus*. While in the first case the quality of the environment decreases, in the second case the same amount of pollution is more detrimental, as agents are more concerned with environmental quality.

As there are no transitional dynamics, this economy is always characterized by (10) and (11). Therefore, by solving the integral in (5), we can obtain $\varphi(K_t)$ as follows

$$\varphi(K_t) = LK_t^{(1-\alpha\mu)(1-\sigma)} a_1^{\mu(1-\sigma)} - \frac{1}{\rho(1-\sigma)} \tag{12}$$

where

$$L = A^{-\alpha\mu(1-\sigma)} \frac{\left[\frac{\sigma-1}{\sigma}A + \frac{\rho}{\sigma(1-\alpha\mu)}\right]^{-\sigma}}{(1-\alpha\mu)(1-\sigma)}. \tag{13}$$

As $[((\sigma-1)/\sigma)A + (\rho/(\sigma(1-\alpha\mu))] = C/K$ is positive, the sign of (13) depends upon the value of σ. For $0 < \sigma < 1$, L is positive, while for $\sigma > 1$, it is negative. Thus, both $\partial\varphi/\partial a$ and $\partial\varphi/\partial K$ are positive, independently of the value of σ. However, as $\partial^2\varphi/\partial a\partial K \gtreqless 0 \Longleftrightarrow \sigma \lesseqgtr 1$, the benefit from adoption increases with the stock of capital when $\sigma < 1$, and it decreases for $\sigma > 1$. In fact, when $\sigma < 1$, the marginal utility of environmental quality increases with consumption, implying that countries with higher capital stocks and levels of consumption derive higher utility from an improvement in environmental quality. In this case, the benefits from adoption increase with the stock of capital. In contrast, for $\sigma > 1$, the marginal utility of environmental quality decreases with consumption, implying that countries with higher capital stocks and levels of consumption derive lower utility from an improvement in environmental quality. Therefore, the benefits from adoption decrease with the stock of capital. The implications of these results are discussed in Section 3.3.

3.2. ANTICIPATED ADOPTION

Using the results from the previous section, we study the effects on optimal growth of an anticipated adoption at a given date T. The problem can be stated as follows:

$$\underset{\{C_t\}}{\text{Max}} \int_0^T \frac{(C_t Q_t^{\mu})^{1-\sigma} - 1}{1-\sigma} e^{-\rho t} dt + e^{-\rho t}\varphi(K_{T_+}) - e^{-\rho T}(\gamma\upsilon + f)$$

s.t.

$$\varphi(K_{T_+}) = L(\beta K_{T_-})^{(1-\alpha\mu)(1-\sigma)} a_1^{\mu(1-\sigma)} - \frac{1}{\rho(1-\sigma)} \tag{14}$$

$$\dot{K}_t = AK_t - C_t$$
$$K_0 \text{ given} \tag{15}$$
$$Q_t = a_0(AK_t)^{-\alpha}$$

where $\varphi(K_{T_+})$ results from (12), taking into account that at the moment of adoption a part of the stock of capital is lost, and the technology changes.

By solving the problem taking T as given, we again obtain conditions (7) and (8). We can use these conditions and (2) to write the optimal path for this economy until the moment of adoption in terms of the ratio C/K. Then, we obtain

$$\frac{\partial(C/K)}{\partial t} = (1-\alpha\mu)\left(\frac{C_t}{K_t}\right)^2 - \Lambda\frac{C_t}{K_t} \tag{16}$$

where

$$\Lambda = \left[\frac{\rho}{\sigma} - \frac{(1-\sigma)(1-\alpha\mu)}{\sigma}A\right] \tag{17}$$

To satisfy the transversality on the stock of capital (9), Λ has to be positive. Moreover, the following transversality condition has to hold[11]

$$\lambda_{T_-} = \frac{\partial\varphi}{\partial K_{T_+}}\frac{\partial K_{T_+}}{\partial K_{T_-}} = \beta\lambda_{T_+} \tag{18}$$

implying that

$$\lambda_{T_-}K_{T_-} = \lambda_{T_+}K_{T_+}$$

Thus, the shadow price of the capital stock must be continuous after taking into account the capital loss due to adoption, to rule out unexploited arbitrage conditions.[12]

Evaluating (7) at $t = T_-$, and using (18), we obtain[13]

$$\frac{C_{T_-}}{K_{T_-}} = \left[\frac{\sigma-1}{\sigma}A + \frac{\rho}{\sigma(1-\alpha\mu)}\right]\delta^{-\frac{(1-\sigma)}{\sigma}} \tag{19}$$

where $\delta = \beta^{(1-\alpha\mu)}(1+\upsilon)^{\mu}$ represents the net effect of both the loss on capital (through β) and the improvement on technology (measured by υ) when adoption takes place.[14] Therefore, δ weights the negative effect of "creative destruction" and the positive environmental impact due to technological change.

The solution to the above differential equation (16), taking the terminal condition (19) into account, implies that before adoption, along the optimal path we have[15]

$$\frac{C_t}{K_t} = \frac{\frac{\sigma-1}{\sigma}A + \frac{\rho}{\sigma(1-\alpha\mu)}}{1 + e^{-\Lambda(T-t)}\left[\delta^{\frac{(1-\sigma)}{\sigma}} - 1\right]} \tag{20}$$

Notice that (10) still characterizes the behavior of the economy after adoption.

We can write the utility of the representative agent as a function of C/K and K. Then, we substitute the optimal C/K given in (20) on the expression of the rate of growth of capital to obtain the optimal path for the stock of capital. This is shown in the Appendix. Thus, the welfare of the representative agent as a function of the time of adoption, T, is given as follows

$$V(T) = a_0^{\mu(1-\sigma)}K_0^{(1-\alpha\mu)(1-\sigma)}L\left[1 + e^{-\Lambda T}\left[\delta^{\frac{(1-\sigma)}{\sigma}} - 1\right]\right]^{\sigma} - \frac{1}{\rho(1-\sigma)} - e^{-\rho T}(\gamma\upsilon + f) \tag{21}$$

and $e^{-\rho T}(\gamma \upsilon + f)$ represents the present value of the costs of adoption. Also,

$$V(0) = a_0^{\mu(1-\sigma)} K_0^{(1-\alpha\mu)(1-\sigma)} L\delta^{(1-\sigma)} - \frac{1}{\rho(1-\sigma)} - (\gamma\upsilon + f) \qquad (22)$$

corresponds to immediate adoption, while

$$V(\infty) = a_0^{\mu(1-\sigma)} K_0^{(1-\alpha\mu)(1-\sigma)} L - \frac{1}{\rho(1-\sigma)}$$

corresponds to no adoption.

3.3. THE OPTIMAL T

The optimal timing of adoption, T^*, is the value of T that maximizes welfare, or (21). The decision about when to adopt depends on the behavior of $V(T)$ with respect to T. The study of $V(T)$ is undertaken in the next propositions.

Proposition 1. *If $\delta < 1$, then $V(T)$ is increasing in T, and the country never adopts, or, $T^* = \infty$.*

Proof. In the Appendix.

If the decrease on capital is so large (small β) that it dominates the improvement on technology (measured by υ), that is, if $\delta < 1$, then it is better to never adopt, as the benefit from adoption is negative. Therefore, a necessary condition for the benefit of adoption to be positive is that $\delta > 1$. Notice that even if there were no other adoption costs ($\gamma = f = 0$), as would occur if the new technology were offered freely, the country may choose not to adopt.

Proposition 2. *Let $\delta > 1$. If (i) $\sigma < 1$, it is always better to adopt at a finite period T than to never adopt. In particular, it is optimal to adopt immediately iff $\partial V/\partial T < 0$, for $T = 0$. For (ii) $\sigma > 1$, the country either adopts immediately or never adopts.*

Proof. In the Appendix.

In Figures 1 and 2 below $V(T)$ is plotted, illustrating the first part of Proposition 2.

If there is an interior solution for T, we may obtain it by maximizing (21) with respect to T. The corresponding first-order condition, $\partial V(T)/\partial T = 0$, is given by

$$\frac{a_0^{\mu(1-\sigma)} K_0^{(1-\alpha\mu)(1-\sigma)} L\sigma \left[\delta^{\frac{(1-\sigma)}{\sigma}} - 1\right] \Lambda e^{-\Lambda T^*}}{\left[1 + e^{-\Lambda T^*}\left[\delta^{\frac{(1-\sigma)}{\sigma}} - 1\right]\right]^{1-\sigma}} = \rho e^{-\rho T^*}(\gamma\upsilon + f) \qquad (23)$$

where Λ is defined in (17). That is, the marginal benefit of adopting at T^* has to be equal to the marginal cost, and both depend on T. Focusing on an interior solution,

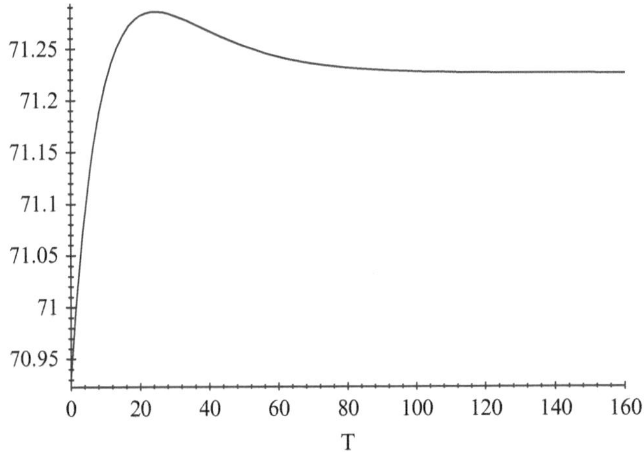

Figure 1. $V(T)$, where $(\partial V(0)/\partial T) > 0, 0 < T^* < \infty, \sigma < 1$.

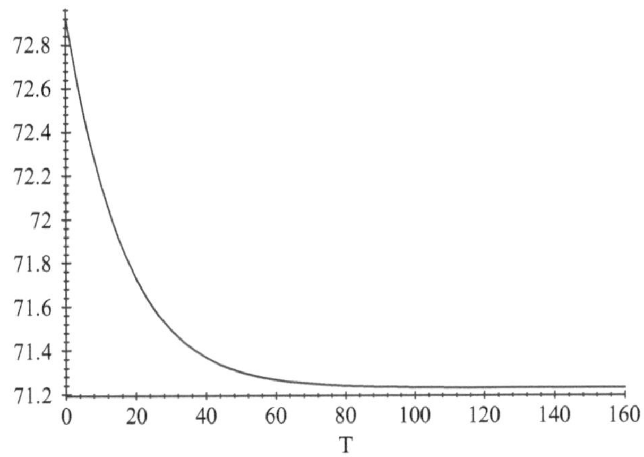

Figure 2. $V(T)$, where $(\partial V(0)/\partial T) < 0, T^* = 0, \sigma < 1$.

we use (23) to determine the capital level at which adoption optimally takes place and to do some comparative statics. These results are stated in Proposition 3.

Proposition 3. *Given the results in Proposition 2, for $\delta > 1$ and $\sigma < 1$, and $0 < T^* < \infty$, the capital level at the optimal timing of adoption is given by*

$$K^*_{T^*_-} = a_0^{\frac{-\mu}{(1-\alpha\mu)}} \left[L^{-1} \frac{\rho(\gamma\upsilon + f)\delta^{\frac{(1-\sigma)^2}{\sigma}}}{\sigma\left[\delta^{\frac{(1-\sigma)}{\sigma}} - 1\right]\Lambda} \right]^{\frac{1}{(1-\sigma)(1-\alpha\mu)}} \tag{24}$$

and it is independent of the initial condition on the stock, K_0. It depends positively on the fixed cost, f, as well as on the quality upgrading, υ, and the marginal cost of adoption, γ, but negatively on the technological quality a_0.

Proof. In the Appendix.

The results in Proposition 2 depend upon the value of σ. If $\sigma > 1$, a "sufficiently high" cost results in a "never adoption equilibrium", while if $\sigma < 1$, a higher adoption cost results in a delay of adoption. The present value of both the benefits and the costs from adoption decreases with T. The decision to either delaying adoption or adopting immediately depends upon which of the two decreases faster. According to our formulation, costs decrease at the discount rate ρ.[16] The rate at which benefits change with T depends upon the behavior of the marginal utility of environmental quality with respect to consumption. When $\sigma < 1$, the marginal utility of environmental quality increases with consumption, and this implies that the benefit from adoption increases with K. Through this effect, the benefit from adoption would increase with T. Therefore, in this case, benefits from adoption will decrease at a rate smaller than ρ. For $\sigma > 1$, instead, the marginal utility of environmental quality decreases with consumption, and we have that the benefit from adoption decreases with K, reinforcing the discounting effect, and implying that the benefit decreases with T at a rate greater than ρ.

Therefore, we obtain that when $\sigma > 1$ and the benefits from adoption decrease with the stock of capital, delaying adoption is never optimal. If the adoption costs are "too high" the country never adopts. Evidence shows that institutions are less efficient in developing countries. This can be captured by higher fixed costs. As poor countries have typically high costs of adoption due to a less efficient institutional framework, adoption may never occur. In order to overcome the "institutional trap", developing countries will have to invest in improving their institutional framework. By subsidizing adoption costs in developing countries, developed countries may contribute to earlier adoption of cleaner technologies in developing countries.

In contrast, for $\sigma < 1$, as the benefits from adoption increase with the stock of capital, it is optimal for the economy to adopt only when it has enough capital. According to the results in Proposition 3, namely that $K^*_{\underline{T}*}$ does not depend upon the initial conditions on K, every country will adopt at the same capital level, or at the same income level. However, depending on the initial conditions on the stock of capital, it may take more or less time to attain that capital level. Thus, poor countries will take longer to attain K^*_{T*}, and therefore, to adopt. This highlights the importance of capital flows between between rich and poor countries. In order to accelerate the process, developed countries may provide compensation to developing ones.

Despite the fact that K^*_{T*} does not depend on initial conditions on capital, we show that country-specific characteristics may lead to different K^*_{T*} between countries. As mentioned above, the institutional framework of the country may determine high costs of adoption, as in the case of high barriers to adoption. As high fixed costs contribute to increasing K^*_{T*}, developing countries may have higher capital thresholds. Therefore, the better the institutional framework of the country, the greater the impact of capital flows from developed countries to developing ones.

Our results are related to those derived in Brock and Taylor (2003), who study investment in abatement, and obtain that initial conditions on the stock of capital do not matter for $K^*_{\bar{T}*}$, for the case of a flow pollutant. As in Brock and Taylor (2004b), we show that country-specific characteristics may induce different capital thresholds for different countries.[17]

Also, Khan and Ravikumar (2002) study the optimal timing of adoption in an environment with capital accumulation where there is a fixed cost of adoption, measured in units of capital, but without externalities. In this context, they also find that it may be optimal to delay adoption.

In our paper, the decision about the optimal timing of adoption involves anticipation of adoption. The consequences of this fact will be shown in the next section.

4. Environmental Quality Prior to Adoption

The behavior of C/K given in (20) for the case in which adoption is anticipated is plotted in Figure 3 below, and its implications are summarized in Proposition 4.

Proposition 4. *For $\delta > 1$, $\sigma < 1$, and $0 < T < \infty$ when adoption is anticipated, prior to technology adoption, growth of consumption and capital accelerates. Moreover, environmental quality decreases.*

Proof. Given (20), if $T \to \infty$, which is equivalent to never adopting, we obtain what we had before, with no adoption. However, if T is finite the anticipation of adoption decreases C/K at each moment, implying an increase in the rate of growth of capital given by (2), prior to adoption. Thus, prior to adoption, as capital is accumulating faster, environmental quality will decrease. The larger the T, the smaller is this effect. Also, from (16), $\dot{C}/C = -\alpha\mu(CK) + ((1 - \alpha\mu(1 - \sigma))/\sigma)A - (\rho/\sigma)$. As it varies negatively with respect to C/K, it also increases.

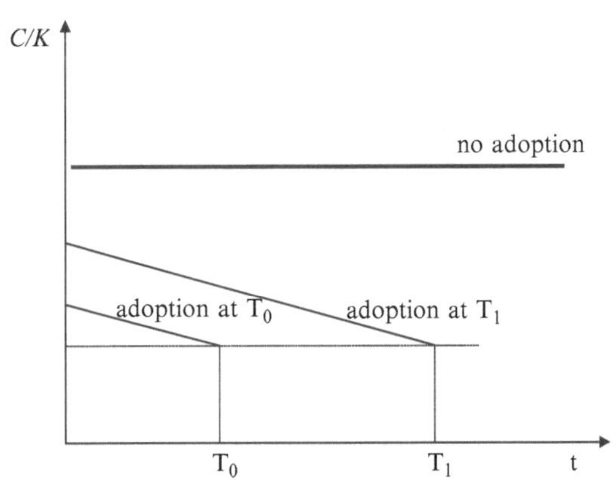

Figure 3. Evolution of C/K prior to adoption, for $\delta > 1, \sigma < 1$.

Therefore, the growth rate of a country that anticipates adoption accelerates before adoption. Despite the fact that the country will adopt a cleaner technology in the near future, capital is accumulating faster, implying that environmental quality will decrease. When the economy anticipates adoption of a cleaner technology at T^* that will be applied to the remaining capital βK, the current benefit from accumulating capital before adoption increases. In fact, this capital will be used after T^* to produce output with lower environmental externalities,[18] also implying that a lower relative cost of consumption is anticipated. In contrast to the results in Khan and Ravikumar (2002), where $\dot{C}/C$ is constant, as there is no environmental externality and, therefore, no concern for the environment (i.e., $\mu = 0$), the optimal growth rate of consumption increases prior to adoption.

5. Conclusion

This paper studies the optimal timing of adoption of a cleaner technology, in the case of a flow pollutant. We show that the adoption decision depends upon the behavior of benefits and costs of adoption over time, which is determined by the nature of the costs of adoption and the behavior of the marginal utility of environmental quality with respect to consumption.

In general, our results may contribute to better understand why developing countries are less willing to adopt cleaner technologies. The implications of these results to policy makers in developed and developing countries caring about the adoption of cleaner technologies are examined.

For a marginal utility of environmental quality increasing with consumption, the benefits from adoption increase with the stock of capital, and delaying adoption can be optimal. Thus, poor countries may take longer to adopt, *ceteris paribus*. Thus, if the developed world is interested in increasing the pace towards the use of green technologies, they may need to give incentives, such as facilitating Foreign Direct Investment in economies that adopt cleaner technologies, or compensate them. In contrast, when the marginal utility of environmental quality decreases with consumption, then the benefits from adoption decrease with the stock of capital, and the countries prefer to adopt immediately, unless costs are so high that the country never adopts. Therefore, a decrease in adoption costs may change the equilibrium from never adopting to immediate adoption. Also, in general, the offer of a cleaner technology may not be a strong enough incentive for adoption if the technology upgrading is not high enough, that is, if the negative effect of "creative destruction" prevails over the positive environmental impact due to technological change.

We also show that country-specific characteristics, as the institutional framework that determines different adoption costs, may generate different adoption decisions. If the marginal utility of environmental quality decreases with consumption, then a reduction in adoption costs may change the optimum from never adopting to immediate adoption. Subsidization of adoption costs by developed countries may play an important role in this context. If the marginal utility of environmental quality increases with consumption, then a reduction in the adoption costs changes the capital threshold, implying an earlier adoption.

Finally, we also obtain that the growth rate of a country that anticipates adoption accelerates before adoption. Thus, prior to adoption environmental quality decreases. The fact that the capital remaining after adoption is operated with the new technology, increases the current benefit from accumulating capital before adoption. The anticipation of a lower relative cost of consumption also contributes to this result.

This paper addresses the problem of a costly adoption from the point of view of a closed economy. An interesting extension would be to consider the role of international trade. Other extensions could be considered, namely, the case of a stock pollutant, and different assumptions on technology as a vintage model. This is left for future research.

Acknowledgments

We are grateful to two anonymous referees and the editor Sjak Smulders for helpful comments. Financial support from NOVAFORUM and FCT is also gratefully ackowledged.

Notes

1. See Bovenberg and Smulders (1995), Elbasha and Roe (1996), and Reis (2001), among others. For a review of the role of adoption in economic growth see Greenwood and Jovanovic (2001).
2. As Romer discusses in Romer (1994) "... there must also be some fixed cost associated with the introduction of each new good. Otherwise, every valuable good would already be in use everywhere." The empirical study of Teece (1977) concludes that "The resources required to transfer technology internationally are considerable."
3. Higher costs may be due to the existence of barriers to technology adoption, especially related to the institutional framework of the country, as in Parente (1994) or Chimeli and Braden (2002).
4. See Oxley (2002).
5. The value of the inverse of the elasticity of intertemporal substitution, σ, is an empirical problem. As mentioned in Balvers and Bergstrand (1997), estimates of σ typically range between 0 and 2. In Issler and Piqueira (2000), an empirical comparative study for Brazil and the USA shows how the values of σ are sensitive to the data used (i.e., annual or seasonally adjusted), and the estimated values of this parameter are consistently lower in the USA than in Brazil.
6. Michel and Rotillon (1995) characterize the first case as "The Compensation Effect", since the consumption desire rises with pollution, while the second case is denoted by "The Distaste Effect", as it reflects some "distaste" of pollution on consumption.
7. We can define pollution as the inverse of the environmental quality Q. If $0 < \alpha \leq 1$, the flow of pollution is never more than proportional to total production. We could have also considered $\alpha \geq 1$, as long as $\alpha\mu < 1$.
8. This is an optimal control problem with an exogenous jump in the state variable at the time adoption takes place, which is also a decision variable. See Seierstad and Sydsaeter (1987), p. 194.
9. Khan and Ravikumar (2002) consider a fixed cost in units of capital, which measures the cost of learning the new method of production. We consider a different specification where the adoption

cost is measured in utility units, which may be understood as an approximation to theirs. We discuss in Section 3 the implications of our cost specification.

10. Notice that while $V(K)$ incorporates the decision on the optimal timing of adoption, $\varphi(K)$ does not.
11. See the Appendix for an alternative approach, following Seierstad and Sydsaeter (1987), p. 194.
12. We are grateful to the editor for suggesting this interpretation.
13. We assume that we always have $(C/K) < A$
14. We will show in Section 3.3 that the decrease on capital has to be smaller than the improvement on technology, that is, $\delta > 1$ is a necessary condition for adoption.
15. The general solution of (16), that is, $(\partial(C/K))/\partial t = M(C_t/K_t)^2 - \Lambda(C_t/K_t)$ is given by $C_t/K_t = \Lambda/(M + e^{\Lambda t} D\Lambda)$, where D represents the constant of integration.
16. Notice that if the fixed cost was introduced in the model as a resource cost, measured in units of K, then it would decrease even in current value terms, as the opportunity cost of capital decreases over time.
17. In the tradition of the Solow model, optimization is absent in Brock and Taylor (2004b). This is not the case in this paper. Thus, our setup is closer to that in Brock and Taylor (2003). See also Brock and Taylor (2004a) for a review of theory and empirics on economic growth and the environment.
18. This result is related to the assumption that the capital remaining after adoption, old and new, will be operated with a higher quality. Alternatively, we could have vintages. However, as a part of the capital is lost after adoption, we already account for some "creative destruction".

References

Balvers, R. J. and J. H. Bergstrand (1997), 'Equilibrium Real Exchange Rates: Closed-Form Theoretical Solutions and Some Empirical Evidence', *Journal of International Money and Finance* **16**(3), 345–366.

Bovenberg, A. L. and S. Smulders (1995), 'Environmental Quality and Pollution-Augmenting Technological Change in a Two-sector Endogenous Growth Model', *Journal of Public Economics* **57**, 369–391.

Brock, W. and M. S. Taylor (2003), 'The Kindergarten Rule of Sustainable Growth', NBER Working-Paper # 9597.

Brock, W. and M. S. Taylor (2004a), 'Economic Growth and the Environment: A Review of Theory and Empirics', NBER Working-Paper # 10854.

Brock, W. and M. S. Taylor (2004b), 'The Green Solow Model', NBER Working-Paper # 10557.

Chimeli, A. and J. Braden (2002), 'Environmental Quality, Environmental Protection and Technology Adoption', University of Illinois at Urbana-Champaign, mimeo.

Dixit, A. and R. Pindyck (1994), *Investment Under Uncertainty*, Princeton NJ: Princeton University Press.

Elbasha, E. H. and T. L. Roe (1996), 'On Endogenous Growth: The Implications of Environmental Externalities', *Journal of Environmental Economics Management* **31**, 240–268.

Greenwood, J. and B. Jovanovic (2001), 'Accounting for Growth', in E. Dean, M. Harper and C. Hulten, eds., *New Directions in Productivity Analysis, NBER Studies in Income and Wealth*, Vol. 63. Chicago: Chicago University Press.

Hall, B. H. and B. Khan (2003), 'Adoption of New Technology', NBER Working-Paper # 9730.

Issler, J. V. and N. S. Piqueira (2000), 'Estimating Relative Risk Aversion, the Discount Rate, and the Intertemporal Elasticity of Substitution in Consumption for Brazil Using Three Types of Utility Function', *Brazilian Review of Econometrics* **20**(2), 201–239.

Khan, A. and B. Ravikumar (2002), 'Costly Technology Adoption and Capital Accumulation', *Review of Economics Dynamics* **5**, 489–502.

Michel, P. and G. Rotillon (1995), 'Disutility of Pollution and Endogenous Growth', *Environmental and Resource Economics* **6**(3), 279–300.

OECD (2001), *Environmental Goods and Services. The Benefits of Further Trade Liberalization.* Paris: OECD.
Oxley, A. (2002), 'Environment Protection and the WTO', in MorrisJulian, ed., *Sustainable Development: Promoting Progress or Perpetuating Poverty?.* London: Profile Books.
Parente, S. (1994), 'Technology Adoption, Learning by Doing, and Economic Growth', *Journal of Economic Theory* **63**, 346–369.
Reis, A. (2001), 'Endogenous Growth and the Possibility of Eliminating Pollution', *Journal of Environmental Economics and Management* **42**, 360–373.
Romer, P. (1994), 'New Goods, Old Theory, and The Welfare Costs of Trade Restrictions', *Journal of Development Economics* **43**, 5–38.
Seierstad, A. and K. Sydsaeter (1987), 'Optimal Control Theory with Economic Applications', in C. J. Bliss and M. D. Intrilligator, eds., *Advanced Textbooks in Economics*. North-Holland.
Stokey, N. (1998), 'Are There Limits to Growth?', *International Economic Review* **39**, 1–31.
Teece, D. J. (1977), 'Technology Transfer by Multinational Firms: The Resource Cost of Transferring Technological Know-How', *Economic Journal* **87**, 242–261.
UN, Press Release (1999), SG/SM/7237.

Appendix

Determination of V(T)

Derivation of (18)

Following Seierstad and Sydsaeter (1987), p. 196, an additional condition at the jump point has to hold. Given the magnitude of the jump, we may write that

$$K_{T_+} - K_{T_-} = g(K_{T_-}, \beta, T) = (\beta - 1)K_{T_-}$$

Then, at the jump point, we have

$$\lambda_{T_+} - \lambda_{T_-} = -\lambda_{T_+} \frac{\partial g(K_{T_-}, \beta, T)}{\partial K_{T_-}}$$

that is,

$$\lambda_{T_+} - \lambda_{T_-} = -\lambda_{T_+}(\beta - 1)$$

which is equivalent to (18).

Expression for K_t.
Utility can be represented by a function of K_t and C_t/K_t, that is,

$$\begin{aligned} V &= \int_0^\infty \frac{C_t^{1-\sigma}(AK_t)^{-\alpha\mu(1-\sigma)}a^{\mu(1-\sigma)} - 1}{1-\sigma} e^{-\rho t} dt \\ &= \int_0^\infty \frac{(C_t/K_t)^{1-\sigma} K_t^{(1-\alpha\mu)(1-\sigma)} A^{-\alpha\mu(1-\sigma)} a^{\mu(1-\sigma)} - 1}{1-\sigma} e^{-\rho t} dt \end{aligned}$$

From (2), and substituting (20), we may write the level of K_t at any moment before adoption as

$$K_t = K_0 e^{\int_0^t \left(A - \frac{\frac{\sigma-1}{\sigma}A + \frac{\rho}{\sigma(1-\alpha\mu)}}{1+e^{-\Lambda(T-\tau)}\left[\delta^{\frac{(1-\sigma)}{\sigma}} - 1\right]} \right) d\tau}$$

Solving the integral we obtain

$$K_t = K_0 e^{At} \left[\frac{1 + e^{-\Lambda(T-t)}\left[\delta^{\frac{(1-\sigma)}{\sigma}} - 1\right]}{e^{-\Lambda(T-t)}\left[\delta^{\frac{(1-\sigma)}{\sigma}} - 1\right] + e^{\Lambda t}} \right]^{1/(1-\alpha\mu)} \tag{A1}$$

Notice that the primitive of $\frac{\frac{\sigma-1}{\sigma}A+\frac{\rho}{\sigma(1-\alpha\mu)}}{1+e^{-\Lambda(T-\tau)}\left[\delta^{\frac{(1-\sigma)}{\sigma}}-1\right]}$ is given by $\left[-\frac{1}{1-\alpha\mu}\ln\left(e^{-\Lambda T}\left[\delta^{\frac{(1-\sigma)}{\sigma}}-1\right]\right.\right.$ $\left.\left.+e^{-\Lambda\tau}\right)\right]$. Thus, with adoption at T, total welfare is given by

$$V(T) = \int_0^T \frac{(C_t/K_t)^{1-\sigma}K_t^{(1-\alpha\mu)(1-\sigma)}A^{-\alpha\mu(1-\sigma)}a_0^{\mu(1-\sigma)}-1}{1-\sigma}e^{-\rho t}dt$$
$$+\int_T^\infty \frac{(C_t/K_t)^{1-\sigma}K_t^{(1-\alpha\mu)(1-\sigma)}(A)^{-\alpha\mu(1-\sigma)}a_1^{\mu(1-\sigma)}-1}{1-\sigma}e^{-\rho(t-T)}dt$$
$$-e^{-\rho T}(\gamma v+f)$$

where K_t and C_t/K_t in the first integral must be substituted by the expressions in (A1) and (20), respectively. The second integral is given by $\varphi(K_{T_+})$ where $K_{T_+} = \beta K_{T_-}$ and K_{T_-} is given by (A1). After some algebra we obtain expression (21).

Proposition 1.

Proof. The derivative of $V(T)$ is $\frac{\partial V}{\partial T} = a_0^{\mu(1-\sigma)}K_0^{(1-\alpha\mu)(1-\sigma)}L\Gamma + e^{-\rho T}\rho(\gamma v+f)$, where Γ is defined as follows

$$\Gamma = \sigma\left[1+e^{-\Lambda T}\left[\delta^{\frac{(1-\sigma)}{\sigma}}-1\right]\right]^{\sigma-1}e^{-\Lambda T}(-\Lambda)\left[\delta^{\frac{(1-\sigma)}{\sigma}}-1\right]$$

Recall that $\Lambda > 0$ to satisfy the transversality condition on capital (9), and that $\left[1+e^{-\Lambda T}\left[\delta^{\frac{(1-\sigma)}{\sigma}}-1\right]\right] > 0$.

For $\delta < 1$ and

(i) $\sigma < 1 : L > 0$ and $\Gamma > 0 \Longrightarrow \frac{\partial V(T)}{\partial T} > 0$.

(ii) $\sigma > 1 : L < 0$ and $\Gamma < 0 \Longrightarrow \frac{\partial V(T)}{\partial T} > 0$.

Therefore, the country never adopts. □

Proposition 2.

Proof. The derivative of $V(T)$ is $\frac{\partial V}{\partial T} = e^{-\rho T}F(T)$ where

$$F(T) = a_0^{\mu(1-\sigma)}K_0^{(1-\alpha\mu)(1-\sigma)}L\Gamma e^{\rho T} + \rho(\gamma v+f)$$

and Γ is defined above. Also, recall that costs decrease at the rate ρ in present value terms. Let $\delta > 1$. In this case, $\frac{\partial V}{\partial T}$ cannot be unambiguously signed, as the first term in $F(T)$ is negative and the second is positive. Thus, we look at the behavior of $\frac{\partial^2 V}{\partial T^2} = e^{-\rho T}\left(\frac{\partial F}{\partial T} - \rho F(T)\right)$. Moreover,

$$\frac{\partial F}{\partial T} = e^{\rho T}\Gamma a_0^{\mu(1-\sigma)}K_0^{(1-\alpha\mu)(1-\sigma)}L\frac{1-\sigma}{\sigma}$$
$$\left[((1-\alpha\mu)A-\rho)+\sigma\Lambda\frac{e^{-\Lambda T}\left(\delta^{\frac{(1-\sigma)}{\sigma}}-1\right)}{1+e^{-\Lambda T}\left(\delta^{\frac{(1-\sigma)}{\sigma}}-1\right)}\right]$$

where $((1-\alpha\mu)A-\rho) > 0$, $\Lambda > 0$ and $\left[1+e^{-\Lambda T}\left[\delta^{\frac{(1-\sigma)}{\sigma}}-1\right]\right] > 0$.

(i) $\sigma < 1 \Rightarrow L > 0, \Gamma < 0$ and $(\delta^{\frac{(1-\sigma)}{\sigma}} - 1) > 0$. Thus, in this case, $\partial F/\partial T < 0$, implying that benefits decrease slower than costs. Also, $F(0)$ can be either positive or negative and $\lim_{T\to+\infty} F(T) = -\infty$. Thus, $\partial V/\partial T$ may be positive or negative for small values of T but it is always negative for large enough T and goes to 0 as $T \to +\infty$. Therefore, we conclude that:

(a) if $\partial V/\partial T > 0$ at $T = 0$ this implies that $F(0) > 0$ and $F(T)$ decreases with T. At some $t = T^*$, $F(T^*) = 0 \Rightarrow \frac{\partial V(T^*)}{\partial T} = 0$ and $(\partial^2 V(T^*)/\partial T^2) < 0$, then $\partial V/\partial T$ becomes negative. Thus, $V(T^*)$ is a maximum, as in Figure 1, and delaying is optimal;

(b) if $\partial V/\partial T < 0$ at $T = 0$, then as $F(T)$ decreases with T, $\partial V/\partial T < 0$ for all T, as illustrated in Figure 2, and the country adopts immediately.

(ii) $\sigma > 1 \Rightarrow L < 0, \Gamma > 0$ and $(\delta^{\frac{(1-\sigma)}{\sigma}} - 1) < 0$. Thus, the term outside the square brackets in $\partial F/\partial T$ is positive but the terms inside the square brackets have opposite signs. Recall that $0 < (1 + e^{-\Lambda T}(\delta^{\frac{(1-\sigma)}{\sigma}} - 1)) < 1$, and $(1 - \alpha\mu)A - \rho > 0$. We show that the term inside the square brackets increases with T,

$$\frac{\partial\left[((1-\alpha\mu)A-\rho)+\sigma\Lambda\frac{e^{-\Lambda T}\left(\delta^{\frac{(1-\sigma)}{\sigma}}-1\right)}{1+e^{-\Lambda T}\left(\delta^{\frac{(1-\sigma)}{\sigma}}-1\right)}\right]}{\partial T} = \frac{-\sigma(-\Lambda)^2 e^{-\Lambda T}\left(\delta^{\frac{(1-\sigma)}{\sigma}}-1\right)}{\left[1+e^{-\Lambda T}\left(\delta^{\frac{(1-\sigma)}{\sigma}}-1\right)\right]^2} > 0$$

At $T = 0$, we have

$$\frac{\partial F}{\partial T_{|T=0}} = \left(1 - \delta^{\frac{-(1-\sigma)}{\sigma}}\right) A(1-\alpha\mu)\sigma + (1-\alpha\mu)A - \rho))\delta^{\frac{-(1-\sigma)}{\sigma}}$$

From (19), in order to have $\frac{C}{K} < A$, this last expression has to be positive, implying that $\frac{\partial F}{\partial T} > 0$ for all T. Thus, benefits decrease faster than costs.

$F(0)$ can be either positive or negative and $\lim_{T\to+\infty} F(T) = \rho(\gamma v + f) > 0$. Thus, $\partial V/\partial T$ may be positive or negative for small values of T and goes to 0 as $T \to +\infty$. Therefore, we conclude that:

(a) if $\frac{\partial V}{\partial T} > 0$ at $T = 0$, then it will always be positive, decreasing to zero as $T \to \infty$, and the country will never adopt;

(b) if $\frac{\partial V}{\partial T} < 0$ at $T = 0$, as $F(T)$ increases with T, it will keep increasing. At some $t = T^*, 0 < T^* < \infty$, $F(T^*) = 0$ and $\frac{\partial^2 V(T^*)}{\partial T^2} > 0$, before becoming positive. Therefore, $V(T)$ reaches its minimum at T^*. Therefore, the country adopts immediately if $V(0) > V(\infty)$ or never adopts.

Proposition 3.

Proof. Evaluating (A1) at $t = T^*$ and rearranging terms, we obtain that at T^*,

$$K^*_{T^*_-} = K_0 e^{AT}\left[\frac{e^{-\Lambda T^*}\delta^{\frac{(1-\sigma)}{\sigma}}}{1+e^{-\Lambda T^*}\left[\delta^{\frac{(1-\sigma)}{\sigma}}-1\right]}\right]^{\frac{1}{(1-\alpha\mu)}}$$

$$= K_0\left[\frac{e^{\frac{(1-\sigma)}{\sigma}[(1-\alpha\mu)A-\rho]T^*}\delta^{\frac{(1-\sigma)^2}{\sigma}}}{\left[1+e^{-\Lambda T^*}\left[\delta^{\frac{(1-\sigma)}{\sigma}}-1\right]\right]^{(1-\sigma)}}\right]^{\frac{1}{(1-\sigma)(1-\alpha\mu)}}$$

Let $\delta > 1,\ \sigma < 1$, and $0 < T^* < \infty$. Rearranging (23), we obtain

$$\frac{e^{\frac{(1-\sigma)}{\sigma}[(1-\alpha\mu)A-\rho]T^*}}{\left[1+e^{-\Lambda T^*}\left[\delta^{\frac{(1-\sigma)}{\sigma}}-1\right]\right]^{1-\sigma}} = \frac{\rho(\gamma v+f)}{a_0^{\mu(1-\sigma)}K_0^{(1-\alpha\mu)(1-\sigma)}L\left[\delta^{\frac{(1-\sigma)}{\sigma}}-1\right]\Lambda\sigma}$$

Substituting in the expression for $K^*_{\underline{T}^*}$, we obtain (24).

Moreover, let us denote φ by

$$\varphi = \left[L^{-1}\frac{\rho(\gamma v+f)\delta^{\frac{(1-\sigma)^2}{\sigma}}}{\left[\delta^{\frac{(1-\sigma)}{\sigma}}-1\right]\sigma\Lambda}\right]^{\frac{1}{(1-\sigma)(1-\alpha\mu)}} > 0$$

Then,

$$\frac{\partial K^*_{\underline{T}^*}}{\partial f} = a_0^{\frac{-\mu}{(1-\alpha\mu)}}\varphi\frac{(\gamma v+f)^{-1}}{(1-\sigma)(1-\alpha\mu)} > 0$$

$$\frac{\partial K^*_{\underline{T}^*}}{\partial v} = a_0^{\frac{-\mu}{(1-\alpha\mu)}}\varphi\left[\frac{(\gamma v+f)^{-1}\gamma}{(1-\sigma)(1-\alpha\mu)}\right] > 0$$

$$\frac{\partial K^*_{\underline{T}^*}}{\partial \gamma} = a_0^{\frac{-\mu}{(1-\alpha\mu)}}\varphi\frac{(\gamma v+f)^{-1}v}{(1-\sigma)(1-\alpha\mu)} > 0$$

$$\frac{\partial K^*_{\underline{T}^*}}{\partial a_0} = a_0^{\frac{-\mu}{(1-\alpha\mu)}-1}\left(-\frac{\mu}{(1-\alpha\mu)}\right)\varphi < 0$$

4. Can Environmental Regulations Boost Growth?

ROB HART
SLU, Department of Economics, Box 7013, 750 07 Uppsala, Sweden. Telephone +46 18 67 17 34, fax +46 18 67 35 02, e-mail rob.hart@ekon.slu.se.

Abstract. I develop a simple new growth model to demonstrate a mechanism through which environmental regulations can boost the growth rate of production towards its socially optimal level, a version of the Porter hypothesis. The mechanism is also likely to operate in much more complex economies, although the net effect of regulations will be uncertain in such economies. In the model, growth is driven by researchers striving for monopoly profits. New technologies must compete with the old for market share. They are not only more productive than the old, but also more environmentally friendly. Introduction of technology standards favours new technologies and therefore increases expected returns to research, hence the quantity of research – and the growth rate – in the economy goes up. This may be a social benefit if knowledge, which drives growth, is underprovided due to spillovers.

Key words: endogenous growth, innovation, environment, Schumpeter, Porter hypothesis.

1. Introduction

Is there necessarily a trade-off between environmental quality and growth? There is a perception that we would like the economy to grow faster than it actually does, and environmental regulations are perceived as having the opposite effect, i.e. braking growth. Thus we must choose between growth and environment. But Michael Porter and others (see for instance Porter and van der Linde (1995)) argue that environmental regulations or standards may trigger innovation which partially *or fully* offsets the costs of compliance. That is, environmental regulations may lead to benefits on the production side of the economy, as well as for the environment (the so-called *Porter hypothesis*). The purpose of this paper is to put forward a model of a simple economy in which a version of the Porter hypothesis holds. The model is a "stripped-down" version of the model of Hart (2004), with the advantage that the Porter mechanism is demonstrated more clearly.

I begin by discussing the literature on the provision of knowledge (and hence growth), the provision of environmental quality, and links between them, in particular the trade-off between production growth and environment. First, note that in a market economy there is nothing logically inconsistent in the notion that an exogenous change (such as the introduction of a policy measure) may lead to *both* an increase in the growth rate of production *and* an improvement in environmental quality. Furthermore, in the absence of the change both growth and environment

L. Bretschger and S. Smulders (eds.), Sustainable Resource Use and Economic Dynamics, 53–70.

may be underprovided in the market economy (due to multiple market imperfections), and hence the change may cause utility to increase unambiguously on both counts.

That growth may be underprovided in the market economy has become accepted with the emergence of new growth theory in which growth is based on knowledge (see for instance Romer (1994) and Jones and Williams (1998, 2000)); knowledge is a public good generated through deliberate effort. That environmental quality – also a public good – may be underprovided in a market economy is a foundation-stone of environmental economics. But in both cases the possibility of underprovision due to market failures is not the same as actual underprovision in the regulated economy; one would expect the regulator to take steps to correct such failures.

The correction of market failures, both in the case of knowledge creation and environment, is more easily said than done. I now consider these problems, beginning with the underprovision of knowledge, which can be linked in turn to underprovision of knowledge-generating effort, henceforth *research*. There are a number of theoretical reasons why attempts by a regulator to increase research effort in the economy are hampered, mainly due to information problems. Research is a process of uncertain outcome and difficult to monitor. Thus if an agent is offered money to perform research which it is not otherwise in her best interests to do (she would rather do something else) then she has an incentive to pocket the cash but carry on with her optimal choice of activity. Furthermore, if research is shunned because of its risky nature, an insurance market should develop. However, given such a market the prospective researcher would have an incentive to take out insurance but not do the research, and then claim the insurance payout. See Arrow (1962) for a seminal discussion. Is there any empirical evidence to support claims of a research shortfall? One example is Jones and Williams (1998) who present evidence, using econometric estimates of returns to Research and development (R&D), that observed spending on research is less than a quarter of socially optimal R&D investment.

Meanwhile, efforts to correct for environmental externalities are hampered by the perception that the measures required will reduce company profits and the growth rate of production (and hence also consumption). This is why the Porter hypothesis, which turns this on its head, is so explosive. What theoretical literature can be found on the subject?

The idea that environmental regulations may induce firms to invest in research is analysed in the literature on industrial organization. For instance, Innes and Bial (2002) show that measures to correct production incentives for a new, clean, good also boost research incentives. In this literature the induced research effort is generally seen as a problem, not a solution, because these models include no knowledge spillovers (and no long-run growth) so the quantity of research in the economy is assumed, ceteris paribus, to be optimal. Innes and Bial (2002) are partly motivated by the observation that technological leaders may lobby for *stricter* environmental standards (they cite cases involving fuel content standards in the USA), the explanation being that such regulations raise rivals' costs, an idea attributable to Salop and Scheffman (1983).

The idea that environmental regulation might boost growth in a general-equilibrium framework is relatively unexplored. There are models, such as Bovenberg and Smulders (1995), where environment is a factor in the production function, so feedbacks from a better environment boost future production (growth). However, this is not a Porter-type mechanism, as the effect is not via incentives to innovate. Xepapadeas and de Zeeuw (1999) analyse, in a vintage model, the effects of environmental regulations on the age of the capital stock of a firm or industry. Regulations may induce a switch to newer, more productive, capital, thus mitigating the costs. However, a full Porter-type effect can only be achieved if there are spillovers between firms, as argued by Mohr (2002). Mohr develops a simple model where firms learn from each other about a new technology (learning-by-doing). Hence there are external economies of scale, and no single firm wishes to be the first mover. Under these circumstances, a regulation to force the adoption of the technology may be beneficial for all firms.

Mohr (2002) analyses an economy with a single, exogenously available new technology. However, it could be extended trivially to an economy with new technologies arriving regularly and exogenously as "manna from heaven", in which case dynamic regulations could boost the long-run growth rate by encouraging adoption. Furthermore, it could be extended and transformed into an endogenous growth model by allowing a research sector in the economy, with the number of researchers determined by an arbitrage condition. Faster adoption induced by regulations should give higher expected profits to researchers, hence boosting both the number of researchers and the long-term growth rate.

There exist in the literature at least two endogenous growth models of the "extended-Mohr" type suggested above. First, Ricci (2007), who begins by setting up a vintage model where the vintages differ in environmental friendliness, and shows that an emissions tax shifts the balance of production towards more environmentally friendly vintages (similar to Xepapadeas and de Zeeuw (1999)). He goes on to endogenize both the quantity and quality – environmentally friendliness – of research as affected by environmental regulation. Both quantity and quality are continuous variables and choices are determined by arbitrage conditions. He finds that the emissions tax increases the value of environmental knowledge, thus encouraging environmentally friendly research at the expense of productivity growth. The net effect on the quantity of research and the rate of growth of production is ambiguous, and the model is solved numerically.

The second example is Hart (2004), which is, like Ricci (2007), a Schumpeterian model owing much to Aghion and Howitt (1992, 1998). Here skilled labourers choose between research and production, hence an arbitrage condition must hold and research quantity is a continuous variable. However, the quality of research effort is a discrete variable—research is either "ordinary" or "environmentally friendly". Technologies are discrete, and there are diminishing returns in production based on a given technology, hence this is a vintage model in which new and old vintages compete against one another. Environmental regulations, by favouring new vintages, may boost the quantity of research while simultaneously influencing its quality (making research more environmentally friendly). The model is solved analytically, but is nevertheless complex with a number of unusual features such as diminishing returns to research effort, and fixed costs per producer in the intermediate-good sector.

The model presented here is closely related to Hart (2004), but it is much more focused on the quantity of research effort as affected by environmental regulations. The contribution is to show in a very simple model how environmental regulations, by favouring recent vintages, may lead to an increase in the quantity of research and hence an increase in the growth rate of production in the economy.

The mechanism of the model is as follows. Skilled labour chooses between research and production on different intermediate vintages. (In market equilibrium, all options must give equal expected returns.) There is a single type of research, which generates designs for new vintages which are both more productive and more environmentally friendly – hence market power and profit. Research also adds to general knowledge, allowing future research to generate even better designs, and hence steady-state growth. The single research type is assumed to boost the environmental friendliness of new vintages more than it boosts their productivity, hence total damages go down when new vintages are introduced.

The social planner, in the absence of environmental damages, spreads production labour over all available vintages, exponentially declining with age. But when older vintages are dirtier they may be abandoned. In the laissez-faire economy production on the newest (cleanest) vintage is low because its holder has a monopoly through patent protection. The quantity of research may be either too high or too low, depending in essence on the balance between incomplete appropriation of benefits (research too low), and the researchers' failure to account for the business-stealing effect of their discoveries (research too high). In the regulated economy, a technology standard is used to prevent production from dirty (old) vintages. The resulting dominance of the latest vintage has the dual consequence of lower environmental damages and higher profits to its discoverer. These higher profits make research more attractive.

The imposition of technology standards does not in itself result in a socially optimal allocation of research and productive effort, but it may move the economy towards the optimal allocation if there is de facto research underprovision. Such underprovision may arise, as discussed previously, because of the information problems connected with supporting research Arrow (1962). The technology standard escapes such problems since it is applied based on the observable characteristics of existing vintages.

In Section 2 I present the model, beginning with the fundamentals and going on to the allocation of production labour and the arbitrage condition between production and research. In Section 3 I discuss the policy relevance of the model, and in Section 4 I summarize the conclusions as well as suggesting possible extensions.

2. A Schumpeterian Growth Model with a Porter-Type Mechanism

2.1. THE FUNDAMENTALS OF THE MODEL

The model is based on Hart (2004), owing much in turn to the Schumpeterian growth models of Aghion and Howitt (1992, 1998). Equations 1–3 for the utility, production, and damage functions are identical to Hart (2004), but subsequently the models differ. In common with Hart (2004), dynamic adjustment in the model economy occurs instantly – there is no shift in allocation as pollution increases (as in Grimaud (1999)), rather the economy goes directly to a steady state whatever the initial conditions. This property, which facilitates solution, is reflected in the form of the utility and damage functions (no stock pollution), and the fact that there is no capital accumulation in the model.

The model is set in infinite, continuous time. At time t there is a single final output, Y_t, used only for consumption (there is no capital in the model), and produced by competitive firms using intermediate goods, for which there is an infinite series of designs. There are negative external effects of production, in the form of damages D_t. New designs are both more productive and cleaner.

The social planner's problem is

$$\max U = \int_0^\infty e^{-rt} u_t \, dt, \quad \text{where } u_t = Y_t / D_t^\phi. \tag{1}$$

Here $\phi \in [0, 1)$ is a parameter determining the weight of damages in the utility function, and r is the (constant) rate of time preference. Thus individuals have intertemporally additive, risk-neutral preferences over consumption and environmental damages. Damages are a "normal bad" in that the greater the production, the greater is the marginal benefit of reducing damages. But a proportional increase in both production and damages (a scaling-up of production) does lead to an increase in utility.

Final-good production is competitive, and intermediate goods are the only inputs. In period s, intermediate designs d_i are available, where $i = -\infty, \ldots, s$, and d_s is the most recently discovered design. Intermediate goods are labelled x_j, where $j = s - i$. So j indicates the vintage of a good's design relative to the most recent design, for which $j = 0$. Each good has an associated productivity A_{s-j}, and gives diminishing returns in final-good production. If the number of designs in use is n then production in period s, Y_s, is

$$Y_s = \sum_{j=0}^{n-1} A_{s-j} x_j^\alpha, \qquad 0 < \alpha < 1. \tag{2}$$

The price p_Y is normalized to unity. Note that damages D do not affect production.

Environmental damages are modelled as flow pollution, for simplicity. They arise as external effects of the process of final-good production. In period s, damages are given by

$$D_s = \sum_{j=0}^{n-1} \frac{A_{s-j}}{E_{s-j}} x_j^{\alpha}, \tag{3}$$

where E_{s-j} is the environmental productivity of the associated intermediate good x_j. Thus, if environmental productivity is constant, then damages rise at an equal rate to production — this is if the scale of production increases. But if environmental productivity also increases, then unit damages fall. If environmental productivity increases faster than ordinary productivity, then *total* damages fall (in a steady state with constant labour allocation).

Damages can be reduced in two ways other than simply reducing production: in the short term by shifting labour between vintages, to concentrate labour on those vintages which are most environmentally friendly (lowest unit damages, A_i/E_i), and in the long term by developing new designs the use of which generates lower unit damages. Such new designs are generated by research.

Skilled labour H may engage in either research R or intermediate-good production L: $H = R + L$. There are constant returns to labour L_j dedicated to any one intermediate x_j:

$$x_j = L_j; \qquad L = \sum_{j=0}^{n-1} L_j. \tag{4}$$

The holder of the latest design can exclude others from its use in production, and hence acts as a monopolist. On the other hand, the knowledge inherent in *older* intermediate designs is non-excludable, so their producers are competitive and price equals marginal cost.

Research success is stochastic, with designs arriving in a Poisson process at rate λ (so the total arrival rate is λR). A new design builds on (and thus raises) existing general knowledge, in two dimensions: production general knowledge A, and environmental general knowledge E. Introducing parameters γ^a and γ^e, we have:

$$A_{s+1} = \gamma^a A_s; \quad E_{s+1} = \gamma^e E_s. \tag{5}$$

Thus new technologies are both more productive and cleaner than the old; furthermore, we restrict the parameters such that the introduction of newer vintages leads to a fall in total damages despite increased production, that is $\gamma^e > \gamma^a$. New designs benefit society both in the present and in the future, since future technologies build on the new knowledge. The productivity improvement also benefits the holder of the design (the successful researcher), since the holder has a monopoly and can therefore restrict production and extract profits. On the other hand, in the absence of environmental regulations, the environmental benefits of the new technology are purely external to the successful researcher.

2.2. ALLOCATION OF PRODUCTION LABOUR

In this section I analyse the allocation of production labour between vintages, both for the social planner, in the laissez-faire economy, and in the regulated economy. The allocation is affected by the knowledge embedded in each vintage, and (in the market) by market power and, potentially, regulation. I take a highly intuitive approach, with mathematical derivations provided in the appendix.

The results can be summarized as follows. The social planner, given zero growth in environmental knowledge E, spreads production over all vintages with an exponential decline with age. Given growth in E, old vintages – which now give both lower production and higher damages – are abandoned. In the laissez-faire production market, production from the newest vintage is low, because the holder of the vintage has a monopoly and therefore reduces production in order to raise the market price, thus maximizing profits (see equation 8). But the regulator, to achieve her preferred allocation, imposes a technology standard allowing only the newest (cleanest) technology to be used. Note that the standard corrects for both the environmental externality and monopoly power.

I begin with the planner's problem. The allocation in period s has no effect on the economy in period $s+1$ (following a new discovery), so the problem is a static one in which production is optimized independently during each period. The problem in period s is thus to maximize u_s, where $u_s = Y_s/D_s^{\phi}$ (see equation 1). Older vintages are less productive, but the presence of diminishing returns in production (equation 2) means that this effect alone leads only to a decline in the use of old vintages rather than their abandonment. However, the fact that older vintages are also dirtier (equation 3), and the fact that damages appear in the denominator of the utility function, leads to their abandonment at some stage. I proceed on the assumption that the planner uses only the latest vintage. The condition for this to hold is derived in the appendix. Inserting equations (2) and (3) into equation (1) gives

$$u_s = A_s^{1-\phi} E_s^{\phi} L^{\alpha(1-\phi)}. \tag{6}$$

Now to the market allocation of labour between vintages, beginning with the laissez-faire market. Given diminishing returns in any given vintage, and the productivity ratio between vintages γ^a, it is straightforward to derive that, ceteris paribus, the ratio of production labour on successive vintages is given by

$$L_j/L_{j+1} = (\gamma^a)^{1/(1-\alpha)}. \tag{7}$$

That is, production labour declines exponentially with age, and the decline is more rapid the larger the steps in productivity (γ^a) and the lower the rate of diminishing returns (α close to unity). However, other things are not always equal, because we assume (Section 2.1) that the design of latest vintage is exclusive to its developer, whereas the designs of older vintages are public goods produced by perfectly competitive firms. The market power enjoyed by the holder of the latest vintage allows her to push up the price by holding down production, and hence make profits. The profit-maximizing quantity of labour, relative to labour employed on the second vintage, is given by

$$L_0/L_1 = (\alpha\gamma^a)^{1/(1-\alpha)}. \tag{8}$$

In the regulated economy, factors such as emissions taxes will affect the balance between vintages. However, I follow Mohr (2002) and Hart (2004) in assuming that the desired effect – that production is based exclusively on the leading, cleanest, vintage – is best achieved via a technology standard prohibiting use of the older, dirtier technologies, hence we have $L_0 = L$.

2.3. RESEARCH ARBITRAGE

What is the optimal balance between research and production, bearing in mind the restriction $H = L + R$? I consider in turn the arbitrage conditions for the social planner and the individual researcher in the market. Here I offer intuitive derivations; for mathematical derivations, see the appendix.

The planner's problem is to maximize the present value of expected utility U, as defined in the integral in equation (1). For arbitrage this means that the marginal effect on U of an increase in production labour must be equal to the marginal effect of an increase in research. I now investigate these effects in turn.

Consider the current flow of utility u: from equation (1) it is clear that this adds directly to U. The marginal increase in u due to an increase in production labour is, from equation (6), equal to $\alpha(1-\phi)u/L$. Using $H = L + R$ we can then say that the marginal effect on utility U of a marginal increase in production labour is $\alpha(1-\phi)u/(H-R)$.

Given the arrival rate λ per researcher, the marginal effect on U of an increase in research is equal to λ times the effect on U of a new discovery. Given a current flow u, then the flow following a new discovery is γu, where $\gamma = (\gamma^a)^{1-\phi}(\gamma^e)^{\phi}$. Thus the boost to current utility from the discovery is $(\gamma-1)u$, and the marginal effect of a research increase is $\lambda(\gamma-1)u$. However, the marginal effect on the present value of utility U is equal to the flow value divided by the effective discount rate, given by the rate of time preference minus the growth rate of utility, $\lambda[(\gamma-1)u]/(r-g_u)$, where

$$g_u = \lambda(\gamma-1)R. \tag{9}$$

Hence the planner's arbitrage equation:

$$\frac{\lambda(\gamma-1)}{r-g_u} = \frac{\alpha(1-\phi)}{H-R}. \tag{10}$$

The solution is where 9 and 10 intersect.[1]

Similar reasoning to the above can take us to the market research arbitrage equation, but now instead of U we need to consider production, Y. Since the final-good sector is competitive, then payments to the intermediate sector are αY (see equation 2). Beginning with the simplest case, the regulated market where only the

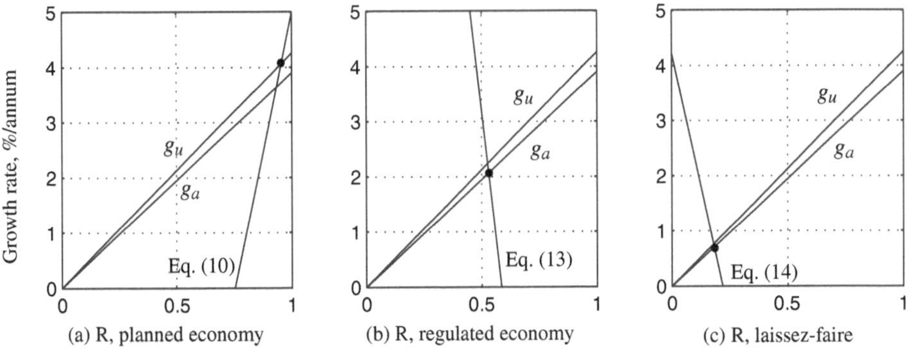

Figure 1. Growth–arbitrage curves for the social planner, regulated market, and laissez-faire market economies. Note that the equilibrium is marked with a dot, and the growth rates of utility and production can be read off given the equilibrium quantity of research.

leading vintage is produced, then the monopoly producer pays labour its marginal revenue product, $\alpha^2 Y/L$. The surplus, $(1-\alpha)\alpha Y$, is captured by the holder of the latest design. The value of the expected income flow from performing research is then $(1-\alpha)\alpha Y \lambda \gamma^a/(r+\lambda R)$, i.e. the value of the flow following a discovery × the arrival rate / the effective discount rate. The effective discount rate is equal to the rate of time preference plus the total arrival rate of new discoveries (since a new discovery destroys the value of the old). Hence research arbitrage with the regulator's solution of only the leading vintage in production is

$$\frac{\lambda\gamma^a}{r+\lambda R}(1-\alpha) = \frac{\alpha}{H-R}. \tag{11}$$

Note that the growth rate of production g_a is, by comparison to equation (9),

$$g_a = \lambda(\gamma^a - 1)R. \tag{12}$$

Substitute into equation (11) to yield

$$\frac{\lambda\gamma^a}{r+g_a/(\gamma^a-1)}(1-\alpha) = \frac{\alpha}{H-R}. \tag{13}$$

Given laissez-faire, the leading producer faces competition from older vintages. Increased production makes wages rise, pushing down the proportion of total payments to the sector which can be captured by the leading producer. The research arbitrage condition is

$$\frac{\lambda\gamma^a}{r+g_a/(\gamma^a-1)}(1-\alpha)L_0/L = \frac{\alpha}{H-R}, \tag{14}$$

where L_0/L is known (see appendix for expression).

In Figure 1 I show the growth and arbitrage equations for the planned, regulated and laissez-faire economies respectively.[2] Note that the growth equations (9) and (12), are identical in each case. Growth increases linearly in research R, hence the positive slope of the growth curves. Growth in utility is somewhat more rapid

Table I. Arbitrage equations in each of the three cases. See equations (9)–(14).

	Planned	Reg	L-F
Appropriation of research benefits	1	$1-\alpha$	$(1-\alpha)L_0/L$
Marginal productivity of research	$\lambda(\gamma-1)$	$\lambda\gamma^a$	$\lambda\gamma^a$
Effective discount rate	$r-g_u$	$r+\frac{g_a}{\gamma^a-1}$	$r+\frac{g_a}{\gamma^a-1}$
Marginal returns to production	$\alpha(1-\phi)/L$	α/L	α/L

than growth in production, since $\gamma^e > \gamma^a$, implying that environmental quality increases over time. The planner's optimum is where the equation for growth in utility (equation 9) crosses the arbitrage equation (10), which can be rewritten as $g_u = r - (H-R)\lambda(\gamma-1)/(\alpha(1-\phi))$. Growth in production can then be read off given the optimum allocation of research labour. On the other hand, the market equilibria are where the equation for growth in production (equation 12) meets the respective arbitrage equations, 14 (laissez-faire economy) and 13 (regulated economy).

2.4. THE PORTER RESULT

Figure 1 illustrates the Porter result. In the planned economy, research is very high and growth rates of utility and production are around 4 per cent. In the regulated market economy, research is considerably lower, and the growth rates are around 2 per cent. Finally, in the laissez-faire economy, research is much lower and growth rates are below 1 per cent. Thus the introduction of regulations banning the use of old, dirty vintages leads to more rapid growth of production in the model economy, and the allocation of labour approaches that chosen by the social planner. Here I analyse the arbitrage equations in detail, in order to better understand the result.

Loosely, we can say that arbitrage is achieved when the appropriability factor multiplied by the marginal productivity of research and divided by the effective discount rate is equal to the marginal returns to production. The terms, in each case, are shown in Table I.

I begin by analysing the difference between the regulated and laissez-faire economies, which lies in the appropriation of research benefits; in laissez-faire, a lower fraction of the social benefits of research are appropriated by the researcher, hence returns to research are lower than in the regulated economy. The reason is that in the regulated economy the owner of the leading design is a monopolist and captures a fraction $1-\alpha$ of social surplus it generates, whereas in the laissez-faire economy the owner faces competition from older vintages and captures only a fraction $(1-\alpha)L_0/L$. Thus the basic Porter result – that the technology standard unambiguously increases research effort by increasing appropriability of research benefits relative to the laissez-faire economy – is unambiguous in the model economy.

The next question is whether a shift in labour allocation towards research is desirable in the model economy; does it move the economy closer to the social planner's allocation? Is research undersupplied in the laissez-faire economy? In this respect the results are ambiguous. I consider each effect in turn.

Firstly, appropriability of research benefits. The planner's appropriability factor is unity, and hence greater than the laissez-faire appropriability, since the planner accounts for society as a whole, which (by definition) appropriates all benefits of research. Hence this effect leads unambiguously to undersupply of research in the laissez-faire market economy.

Turning to the marginal productivity of research, inspection of Table I reveals two differences between the planned economy and the market economies. Firstly, researchers in the market care about the production boost γ^a, whereas the social planner cares about the utility boost γ. (Recall that $\gamma = (\gamma^a)^{1-\phi}(\gamma^e)^{\phi}$.) Since $\gamma^e > \gamma^a$ this effect also leads to undersupply of research in the laissez-faire market economy. Secondly, the planner values the *increase* in utility following a discovery $(\gamma - 1)$, whereas the private researcher values the *level* of productivity (γ^a): researchers do not account for the *business-stealing effect* Aghion and Howitt (1992). This effect is unique in that it shifts the market arbitrage curve to the left relative to the planner's, tending to raise market research effort *above* the socially optimal level. It is important if steps forward are small. Thus it is possible, if steps forward are small, that the Porter effect, boosting research, is undesirable since research is already at or above it's socially optimal level. The intuition is that even a small step ahead of the competition may give a successful researcher a lot of market power, and hence profit; on the other hand, the social benefits may be small, since the successful researcher's gain is largely the previous incumbent's loss.

Now the effective discount rate. It is this term which includes the growth rate, thus completing the growth–arbitrage pairing. For the planner, the discount factor is equal to the *difference between the pure rate of time preference r and the growth rate of utility* – rapid growth makes the present value of future production higher, and therefore makes (future-oriented) research more worthwhile. On the other hand, for the individual agent, the discount factor is equal to the *sum of the rate of time preference and the arrival rate of new discoveries* (a linear function of the growth rate). This is because the arrival of a new discovery destroys the value of the current design. This difference explains why the planner's arbitrage equations have positive slope, whereas the market arbitrage equations have negative slope. It is also, of course, a very important reason why market research effort tends to be below the socially optimal level. This "research deficit" is large when r, the pure rate of time preference, is small relative to λ, the arrival rate of discoveries.

Finally, marginal returns to production, where the only difference is the term $1 - \phi$ in the planner's equation. This depends on the fact that the planner cares about environmental damages, which increase when production effort increases with constant technology, whereas the owner of the leading discovery is indifferent to environmental damages the effects of which are spread over the entire population. Hence this term also leads to lower research effort in the market economy than is socially optimal.

3. The Results and their Robustness

Recall that the aim of the paper is to highlight a mechanism by which environmental technology standards – introduced dynamically to favour new, clean, technologies – may generate higher production growth in the economy, as well as higher discounted utility. In this section I discuss how robust the results are likely to be to changes in the model making it more realistic. First I discuss the results in the model economy and the key features of the model which give rise to them, and then I go on to discuss some limitations to the model and the probable effects of correcting them.

3.1. KEY FEATURES OF THE MODEL

The essential result of the model is that the introduction of standards in a previously laissez-faire market economy unambiguously increases the growth rate of production. It is not unambiguous that this change leads to an increase in utility, but it is very likely to do so since research is likely to be underprovided in the laissez-faire economy.[3]

Despite these clear results, the model should not be interpreted as giving a green light to all sorts of environmental technology standards. On the contrary, it suggests conditions under which such standards may have a positive effect on growth. These conditions include the emergence of new, clean technologies which are also more productive but must compete with older rivals. Furthermore, these technologies should embody knowledge which can be built on in the future rather than leading to a dead end. While this seems entirely reasonable, three further essential features of the model are more open to debate.

Firstly, it is essential to be aware that the Porter-type result is only possible because the standard corrects multiple externalities in a previously laissez-faire economy; while optimally correcting the distribution of production labour for environmental damages in the presence of market power, it simultaneously encourages research. Thus it may be argued that in practice such a standard should be combined with further instruments in order to get closer to an *optimal* allocation in the economy, or even that the standard should be *replaced* by a combination of alternative instruments. The counterargument, in favour of increased stringency in the regulation of old, dirty, technologies (for instance, through dynamic introduction of technology standards), is (i) that there is evidence that the existing panoply of instruments fails in practice to correct for the under-provision of research in the economy (Arrow (1962) and Jones and Williams (1998 and 2000)), and (ii) that stringent regulation or prohibition of dirty technologies has fewer implementation problems associated with it than more conventional means of supporting innovation, since such regulation is based on observable characteristics of technologies, i.e. their relative environmental friendliness.

Secondly, only one type of research is possible in the model, and it leads to more rapid increase in environmental productivity than in labour productivity ($\gamma^e > \gamma^a$). Hence environmental damages fall over time. The idea that higher productivity and lower environmental damages may come hand-in-hand is very natural, as both may be correlated to a reduction in resource use, waste, and physical size (see Porter and van der Linde (1995) for more discussion and examples). However, this relationship is very unlikely to hold generally. If the other extreme were true, that labour productivity always increased more rapidly than environmental productivity, then damages would increase over time despite technology becoming more environmentally friendly. This is of course perfectly possible, as discussed by Mohr (2002). Finally, the more reasonable alternative is that researchers can alter the quality of their research, for instance by making it less productive but more environmentally oriented in response to environmental technology standards. In such a case then the effect of the standards on environmental quality is likely to be enhanced, whereas the effect on production growth becomes ambiguous (see Hart (2004)).

Finally, the independence of the research sector from the intermediate sector is critical to the results, a point which is best understood through consideration of the opposite case. If innovation is controlled by the leading companies in the intermediate sector, then technology standards may lead to a situation with no innovation since these companies prefer the status quo; in such a situation emissions taxes may be a more effective instrument.

3.2. LIMITATIONS OF THE MODEL

Apart from the key features discussed above, the model has a number of limitations which are likely to affect the results to a lesser extent.

A simplification in the Schumpeterian framework is that skilled workers are assumed to be risk-neutral, thus maximizing the expected net present value of income flows. In practice they are likely to be risk-averse, and, following Arrow (1962), this affects the arbitrage condition between research (risky) and production (safe) – nowhere more so than in the sort of Schumpeterian framework set up here, where a single successful researcher (let us call him Bob Stiles) captures the intellectual property rights to the design used in all production. If agents were risk-averse then research would become even less attractive, thus further increasing the need to find ways of encouraging research activity and strengthening the conclusions of the model.

The vintage model is highly simplified in several respects, such as the handling of capital accumulation, learning effects, patent protection, and the balance between vintages.[4] Capital accumulation is excluded; new vintages (once discovered) can be adopted without cost. This simplifies the dynamics but means that the costs of abandoning old vintages are understated. With capital accumulation, both the social planner's optimum and the laissez-faire equilibrium would involve less use of the newest vintages than in the model as it stands. However, it is not clear what the effect would be on the difference between these solutions, hence the benefits of banning old vintages may be just as great in such a setting as they are in the model as it stands. Furthermore, as a referee noted, switching vintages should involve the abandonment of some capital and the accumulation of other capital, processes which may have significant environmental effects. Since the model does not include such processes, it may overstate the benefits of switching rapidly to newer vintages from an environmental standpoint as well. Cunha-e-Sá and Cunha-e-Sá and Reis (Forthcoming) – elsewhere in this volume – analyse the timing of the decision to switch to a new environmentally friendly vintage given capital costs. I am not aware of vintage models which also account for the environmental costs of capital.

Learning effects are also excluded; that is, new technologies do not increase in productivity over time through use. As with capital accumulation, the inclusion of learning effects would affect both the planner's and the laissez-faire solution, and hence the effect of the strength of the "Porter effect" of regulations is not clear. Indeed, Mohr's model Mohr (2002) of a Porter effect through regulating technology adoption actually builds on the existence of learning effects which hinder market adoption.

Patents on old technologies are assumed to be ineffective. This affects the laissez-faire solution alone, leading to lower profits to the discoverers of new designs, both in the first period (due to increased competition from older vintages) and in subsequent periods (when profits are zero due to loss of patent protection). The effect of allowing long-lived patents would then be to boost profits to discoverers in the laissez-faire economy, while leaving the regulated economy unaffected. Hence the Porter effect would be weakened.

Finally, it is assumed that only the latest vintage is used in the social optimum. This simplifies the analysis greatly, but again it leads to an overstatement of the effect that optimal environmental regulation will have on research effort, compared to an economy in which the use of more than one vintage is permitted.

4. Conclusions

I conclude with a brief discussion of possible evidence. Mohr (2002) points out that if the Porter result depends on the correction of a coordination failure between firms, then although individual firms may be against regulations, the industry as a whole should support them. However, the mechanism analysed in this paper depends on the inability of research workers to reap the full social benefits of their discoveries, and those who control the production process may be happy with the status quo rather than risking a more open industry with new technologies being introduced more frequently, since such openness may lead to leadership in the industry changing hands.

On the other hand, independent researchers should support such standards, especially successful ones. This is a trivial conclusion: which researcher, owning the rights to a new design, would not like to see her rivals' designs outlawed? However, a more interesting conclusion slightly beyond the model is that firms with a technological edge over their rivals might be found to support such regulations. This is exactly the observation made by Innes and Bial (2002), who cite cases involving fuel content standards in the USA. Furthermore, they show that measures to correct production incentives for a new, clean, good also boost research incentives. However, they conclude that this is a disadvantage from a socioeconomic viewpoint, as their model does not include knowledge spillovers.

Acknowledgements

Thanks to the Leverhulme Trust (UK) and the Department of Land Economy, Cambridge, for support during the early stages of work on this paper. Thanks also to two anonymous reviewers, Clas Eriksson, and Sjak Smulders for their helpful comments.

Notes

1. If Equations (9) and (10) intersect outside the allowed zone ($L > 0$, $R_e > 0$) the result is a corner solution with either no research, and hence zero growth, or no production, never a reasonable solution.
2. The parameter values are $\lambda = 0.013$, $\gamma^a = 4$, $\gamma^e = 5$, $H = 1$, $r = 0.05$, $\alpha = .3$, $\phi = 0.3$. Thus we have very large steps forward in productivity, consistent with an extremely "coarse-grained" model where only one type of technology is used at a time in the optimum.
3. Recall that research underprovision depends on effects such as knowledge spillovers outweighing the business stealing effect which tends to lead to too much research in the market economy. For further discussion of the likely degree of research under-provision, see Jones and Williams (1998, 2000).
4. For an example of a vintage model with technological progress, capital investment, and learning see Feichtinger et al. (2006).

References

Aghion, P. and P. Howitt (1992), 'A Model of Growth Through Creative Destruction', *Econometrica*, 60:323–51.

Aghion, P. and P. Howitt (1998), *Endogenous Growth Theory*. MIT.

Arrow, K. J. (1962), 'Economic Welfare and the Allocation of Resources for Invention', in R. Nelson, ed., *The Rate and Direction of Inventive Activity*. Princeton NJ: Princeton University Press.

Bovenberg, A. L. and S. Smulders (1995), 'Environmental Quality and Pollution-Augmenting Technological Change in a Two-Sector Endogenous Growth Model', *Journal of Public Economics*, 57:369–391.

Cunha-e-Sá, M. A. and A. B. Reis (2007), 'Optimal Timing of Adoption of a Green Technology', *Environmental and Resource Economics*, 36:35–55.

Feichtinger, G. R. F. Hartl, P. M. Kort, and V. M. Veliov (2006), 'Capital Accumulation Under Technological Progress and Learning: A Vintage Capital Approach', *European Journal of Operational Research*, 172:293–310.

Grimaud, A. (1999), 'Pollution Permits and Sustainable Growth in a Schumpeterian Model', *Journal of Environmental Economics and Management*, 38:249–266.
Hart, R. (2004), 'Growth, Environment and Innovation — A Model with Production Vintages and Environmentally Oriented Research', *Journal of Environmental Economics and Management*, 48(3):1078–1098.
Innes, R. and J. J. Bial (2002), Inducing Innovation in the Environmental Technology of Oligopolistic Firms', *Journal of Industrial Economics*, 50(3):265–287.
Jones, C. I. and J. C. Williams (1998), Measuring the Social Return to R&D', *Quarterly Journal of Economics*, 113:1119–1135.
Jones, C. I. and J. C. Williams (2000), 'Too Much of a Good Thing? The Economics of Investment in R&D', *Journal of Economic Growth*, 5:65–85.
Mohr, R. D. (2002), 'Technical Change, External economies, and the Porter Hypothesis', *Journal of Environmental Economics and Management*, 43(1):158–168.
Porter, M. E. and C. van der Linde (1995), 'Toward a new Conception of the Environment–Competitiveness Relationship', *Journal of Economic Perspectives*, 9(4):97–118.
Ricci, F. (2007), 'Environmental Policy and Growth when Inputs are Differentiated in Pollution Intensity', *Environmental and Resource Economics*, forthcoming.
Romer, P. M. (1994), 'The Origins of Endogenous Growth', *Journal of Economic Perspectives*, 8(1):3–22.
Salop, S. C. and D. T. Scheffman (1983), 'Raising Rivals' Costs', *American Economic Review*, 73(2):267–271.
Xepapadeas, A. and A. de Żeeuw (1999), 'Environmental Policy and Competitiveness: the Porter Hypothesis and the Composition of Capital', *Journal of Environmental Economics and Management*, 37(2):165–182.

Appendix

THE PLANNER'S ALLOCATION BETWEEN VINTAGES

I consider the allocation problem in period $s+1$, to facilitate use of the results in the derivation of the planner's arbitrage condition. From equations (1), (2), and (3), and subject to $\sum_j L_j = L$ (equation 4), the planner's problem is

$$\max_{x_j,n} u_{s+1} = \frac{Y_{s+1}}{D_{s+1}^{\phi}} = \frac{\sum_{j=0}^{n-1} A_{s+1-j} x_j^{\alpha}}{\left(\sum_{j=0}^{n-1} (A_{s+1-j}/E_{s+1-j}) x_j^{\alpha}\right)^{\phi}}. \tag{A1}$$

From equations (4) and (5), the problem can be written

$$\max_{L_j,n} u_{s+1} = A_{s+1}^{1-\phi} E_{s+1}^{\phi} \frac{\sum_{j=0}^{n-1} L_j^{\alpha} (\gamma^a)^{-j}}{\left(\sum_{j=0}^{n-1} L_j^{\alpha} (\gamma^e/\gamma^a)^j\right)^{\phi}}, \tag{A2}$$

subject again to $\sum_j L_j = L$. Older vintages are disfavoured for two reasons: they are less productive (the factor $(\gamma^a)^{-j}$ in the numerator), and they generate more damages per unit of production (the terms in the denominator). When $n = 1$, equation (A2) can be simplified to equation (6).

The condition for $n = 1$ to be the planner's optimal choice is that it gives greater utility than when $n = 2$ (if this is true, then still higher values of n must give even lower utility). The condition is therefore that

$$L^{\alpha(1-\phi)} \geqslant \max_{L_0} (1/\gamma) \frac{\gamma^a L_0^{\alpha} + (L - L_0)^{\alpha}}{((\gamma^a/\gamma^e) L_0^{\alpha} + (L - L_0)^{\alpha})^{\phi}}. \tag{A3}$$

It is straightforward to verify that the condition is satisfied for the parameters chosen in Figure 1.

MARKET ALLOCATION BETWEEN VINTAGES

I initially work out the market allocation over x_j given laissez-faire. I then consider the regulated economy where a technology standard stipulates that only the latest (cleanest) vintage may be used.

Consider the design of vintage j during period s. I define its selling price (to the final-good sector) as $p_{s,j}$, and the wage (to skilled labour) as w_s. Profits are then given by

$$\pi_{s,j} = p_{s,j}x_{s,j} - w_s L_j. \tag{A4}$$

Profit-maximization by final-good producers implies that $p_{Y,j}\partial Y/\partial x_j = p_{s,j}$. Then (using the final-good production function, equation 2) we have

$$\frac{\partial Y}{\partial x_j} = \alpha \frac{A_s}{(\gamma^a)^j} x_j^{-(1-\alpha)} = p_{s,j}. \tag{A5}$$

These are the demand equations for the intermediate producers. The leading good is patent-protected, so we put equation (A5) into the profit equation (A4) and optimize (recalling that $x_j = L_j$—section 2.5), to yield

$$L_0 = \left(\frac{\alpha^2 A_s}{w_s}\right)^{1/(1-\alpha)}. \tag{A6}$$

Putting this back into equation (A4) gives

$$\pi_s = w_s L_0 \frac{1-\alpha}{\alpha}. \tag{A7}$$

What about the remaining (older) vintages, the designs for which are public goods? If these are produced competitively, then (recalling that $x_j = L_j$, and hence using $MC_{s,j} = \partial TC/\partial x_j = w_s$) the condition that $MC_{s,j} = p_{s,j}$ yields

$$L_j = \left(\frac{\alpha \frac{A_s}{(\gamma^a)^j}}{w_s}\right)^{1/(1-\alpha)}. \tag{A8}$$

Now for the allocation between vintages. We can combine equations (A6) and (A8) to give equation (8), whereas for $j > 0$ we have, directly from equation (A8), equation (7).

Finally, given that $L = \sum_j L_j$, it is straightforward to show that

$$L_0/L = \frac{\alpha^{1/(1-\alpha)}\left((\gamma^a)^{1/(1-\alpha)} - 1\right)}{\alpha^{1/(1-\alpha)}\left((\gamma^a)^{1/(1-\alpha)} - 1\right) + 1}. \tag{A9}$$

PLANNER'S RESEARCH ARBITRAGE

I solve for the planner's arbitrage equation by assuming the existence of a steady state, calculating the allocation, and then confirming that the allocation is indeed state-independent (thus justifying the initial steady-state assumption). I assume parameters such that $n = 1$.

I denote the present value of utility by U. We already have an expression for utility in period $s+1$—equation (6). Normalizing the present time to be $t = 0$, I now define the function $\Pi(q, t)$ as the probability that a further q innovations have occurred at time t. Then (given $A_{s+1} = \gamma^a A_s$, $E_{s+1} = \gamma^e E_s$), and defining a new parameter $\gamma = (\gamma^a)^{1-\phi}(\gamma^e)^{\phi}$, expected utility at future time t can be written $u_{s+1}\sum_{q=0}^{\infty}\Pi(q,t)\gamma^q$. Substituting in the optimal value of u_{s+1} (equation 6) and inserting into the planner's dynamic utility maximization problem (equation 1), the planner's problem at $t = 0$, period $s+1$ can be written:

$$\max U_e = \int_0^{\infty} e^{-rt}\gamma A_s^{1-\phi} E_s^{\phi} L^{\alpha(1-\phi)} \left(\sum_{q=0}^{\infty}\Pi(q,t)\gamma^q\right) dt. \tag{A10}$$

From the definition of the Poisson distribution (and given that the total arrival rate in the economy is λR), $\Pi(q,t) = ((\lambda R^t)^q/q!)e^{-\lambda R^t}$. Using the series summation of an exponential ($e^x = 1 + x + x^2/2! + x^3/3! + \cdots$), and the restriction on total labour $H = L + R$, this gives

$$U = \gamma \frac{A_s^{1-\phi} E_s^{\phi} (H - R)^{\alpha(1-\phi)}}{r - \lambda R(\gamma - 1)}. \tag{A11}$$

The first-order condition in research labour gives the arbitrage condition, equation 10. The current state of knowledge (A_t, E_t) does not appear in the arbitrage equation, so we do indeed have a steady-state solution.

MARKET RESEARCH ARBITRAGE

The fundamental arbitrage equation states that the wage must be equal to the expected marginal returns to research, hence

$$w_s = \lambda R_s V_{s+1}, \tag{A12}$$

where V_{s+1} is the value of a discovery, and w_s is the wage in intermediate production. The value of a discovery (given that patent protection runs out after the next discovery) is given by the Bellman equation $rV_{s+1} = \pi_{s+1} - \lambda R_{s+1} V_{s+1}$, hence

$$V_{s+1} = \frac{\pi_{s+1}}{r + \lambda R_{s+1}}. \tag{A13}$$

Inserting equation (A7) (adjusted for period $s + 1$) gives

$$V_{s+1} = \frac{w_{s+1} L_0 (1 - \alpha)/\alpha}{r + \lambda R_{s+1}}. \tag{A14}$$

Recalling the fundamental arbitrage condition (A12), and assuming a steady state so $R_s = R_{s+1} = R$ and $w_s = w_{s+1}/\gamma^a$, gives

$$\frac{w_{s+1}}{\gamma^a} = \lambda \frac{w_{s+1} L_0 (1 - \alpha)/\alpha}{r + \lambda R}. \tag{A15}$$

Given laissez-faire then L_0 is given by equation (A9). Inserting this into equation (A15) gives the laissez-faire research arbitrage condition, equation (14). On the other hand, research arbitrage with the regulator's solution of only the leading vintage in production is given by equation (13).

5. General Purpose Technologies and Energy Policy

ADRIAAN VAN ZON* AND TOBIAS KRONENBERG**

* *UNU-MERIT/Maastricht University, P.O. Box 616, 6200 MD Maastricht, The Netherlands, e-mail: Adriaan.vanZon@MERIT.unimaas.nl*
** *Forschungszentrum Jülich GmbH, Institut für Energieforschung - Systemforschung und Technologische Entwicklung (IEF-STE), D-52425 Jülich, Germany*

Abstract. We employ a general purpose technology model with endogenous stochastic growth to simulate the effects of different energy policy schemes. A Research and Development (R&D) sector produces endogenous growth by developing radical and incremental technologies. These innovations result in blueprints for intermediate goods, which require raw capital and either carbon-based or non-carbon-based fuels. A carbon tax therefore affects not only the final production sector but also the R&D sector by making the development of non-carbon-based technologies more attractive. Due to path dependencies and possible lock-in situations, economic policy can have a significant long-term impact on the energy structure of the economy. We examine the effects of different carbon policies on growth and environmental quality. We find that an anti-carbon policy may reduce growth initially, but in the long run there is a strong potential for a "double dividend" due to faster growth and reduced pollution.

Key words: general purpose technology, carbon tax, R&D, growth, carbon fuel consumption

1. Introduction

The current bad news about the adverse impact of human activity on the environment,[1] stresses the need to reconcile further economic growth with environmental protection. Since the lion's share of environmental damage is caused by the consumption and distribution of energy, a decoupling of growth from pollution will require a massive reorganization of the energy system, possibly leading towards a hydrogen economy (Rifkin 2002). We are interested in finding out how economic policy may help to promote the transition from the current energy system to a sustainable alternative. There have been several energy transitions in the past, all of which were the result of technological breakthroughs. It thus seems reasonable to expect that technological progress will also be a *conditio sine qua non* for future energy transitions.

Technological progress has never been a smooth process. Schumpeter (1939) suggested that drastic innovations may give rise to technological waves, causing long-run cycles in GDP growth. A *drastic innovation* is a radically new idea which is reached after deliberate efforts at combining previously unrelated ideas.[2] Such drastic innovations are a risky enterprise from an economic point of view, because of uncertainty regarding the actual usefulness of the new idea. But from a

L. Bretschger and S. Smulders (eds.), Sustainable Resource Use and Economic Dynamics, 71–99.

technological point of view, the most important characteristic of a drastic innovation is that it opens up new opportunities for further development and expansion of the field of application of this radically new idea. The concept of a drastic innovation is, therefore, closely related to what Bresnahan and Trajtenberg (1995) and Helpman (1998) refer to as a "General Purpose Technology" (GPT). Drastic innovations are usually contrasted with *incremental innovations*. The latter are highly path-dependent, following specific technological trajectories (Dosi 1988) that define, in our terminology, a so-called "technology family" with the drastic innovation serving as its core.[3]

In order to incorporate the crucial features of a GPT into one model, we would have to allow for both vertical innovations (drastic innovations that form the core of a new GPT or technology family) in the sense of Aghion and Howitt (1992,1998a) and horizontal innovations (incremental innovations that raise the productivity of an existing GPT) for which we will largely follow Romer (1990). In both dimensions there will be Love of Variety effects (cf. Romer (1990), van Zon and Yetkiner (2003)), and researchers will have to decide to engage in either basic research that is aimed at finding new cores that form the heart of a new GPT, or in applied R&D that is aimed at expanding the field of application of the core through the addition of "peripheral" innovations. Thus, each successful basic R&D project gives rise to follow-up applied R&D projects.

The hydrogen economy as envisaged by Rifkin (2002) perfectly fits the GPT-bill, as several existing technologies (most notably fuel cells and complex IT systems) must be combined to form a radically different energy distribution system. Hence, if we want to analyse the transition towards a hydrogen economy, it seems only natural to adopt a technology family framework, as outlined above. In such a framework, different technology families should be allowed to coexist at the same time, as it is the case in practice. This has the added bonus of allowing for a relatively smooth transition.

In order to implement this idea, we will use the GPT model by van Zon, Fortune, and Kronenberg (2003), further referred to as the STAGEPOST (Stochastic Arrival of General Purpose Technologies) model, where growth occurs as the result of intentional basic research on the cores of new GPTs and applied research on new peripherals for the further expansion of existing GPTs. The model takes into account the uncertainty associated with drastic innovations by drawing the parameters that determine a technology's characteristics from a random distribution. Depending on these parameters, a new technology may result in a successful GPT with many practical applications, a complete failure (further referred to as "failed" GPTs), or anything in between these two extremes. Developing a new core of a GPT generates technological complementarities in the sense of Carlaw and Lipsey (2001), because the inventor of the core enables other researchers to begin applied R&D in the new technology.

In order to add an energy dimension to the STAGEPOST model, we extend it in several ways. First, we allow for different types of fuels, a carbon-based one and a non-carbon-based one, and assume that carbon-based fuels generate adverse environmental externalities. Second, we introduce two different types of technical progress giving rise to the development of the cores and peripherals of new carbon-based GPTs or non-carbon-based GPTs. Third, we add emissions thus defining (albeit in a very simplistic way) part of the trade-off between growth and environmental quality.

It should be stressed here from the outset that we are primarily interested in finding out how the interaction between different types of R&D activity may influence growth patterns, but also how the intrinsically stochastic nature of these R&D processes may influence the effectiveness of policymaking in the face of path dependencies and lock-in effects. Especially the latter has forced us to formulate a *simulation model* that, apart from its production structure and the way in which the different sectors interact with each other, is as simple as possible. The simulation model is most certainly not a full-fledged completely intertemporally consistent general equilibrium model, but rather a combination of such intertemporal features where needed and more "behaviouristic" features (cf. Solow 2000) where possible.[4]

The extended STAGEPOST model has some features in common with the Matsuyama (1999) model that also generates endogenous cycles in long-run growth. In Matsuyama's model one phase is characterized by capital accumulation, which runs into diminishing returns and thus cannot generate long-run growth. This Solowian phase is followed by a Schumpeterian innovation phase, and the interplay between the two phases generates long-run growth. The major achievement of Matsuyama's model is the integration of neoclassical capital accumulation and Schumpeterian innovation-based growth. The STAGEPOST model takes that integration a step further, but here it is the expansion of GPTs through applied R&D that eventually runs into diminishing returns, thus inducing a shift towards basic R&D, which in turn generates new possibilities for applied R&D. In short, the STAGEPOST model emphasizes the Schumpeterian aspects of growth, but also the role of R&D in both accumulation processes, as well as the uncertainties surrounding these processes, apart from disaggregating R&D activity itself into basic R&D and as many applied R&D processes as there are GPTs.

The extended STAGEPOST model also contributes its own specific features to the literature that are not present in the Matsuyama model. *First*, it introduces asymmetries in the intermediate goods market. These asymmetries arise from our assumption that the contribution of the latest peripherals/intermediates to a GPT decreases as more and more peripherals are added to the GPT.[5] The corresponding asymmetries in profit opportunities in the intermediate goods sectors that implement the ideas produced by the R&D sector, are at the heart of the interplay between basic and applied R&D, where, as Yetkiner (2003) has pointed out, falling profits provide the real incentives for R&D to find the next completely new technology. *Second*, we have multiple GPTs that are active at the same time, as in real life, and as opposed to more standard GPT models. *Third*, we have cyclical growth that does not depend on the reallocation of homogeneous labour between production and R&D activities, as it is the case in the standard GPT approach by Bresnahan and Trajtenberg (1995), or in more recent work focusing on the clustering of R&D activity against a GPT background that explains booms and busts in the business cycle (Francois and Lloyd-Ellis 2003), or in (Matsuyama 1999). Rather than this being a weakness of our model, we feel that it is actually a strong feature, as it is hard to imagine that the "eye-hand-coordination" in the words of Romer (1990) is suited in practice to perform R&D tasks, and the other way around. *Fourth*, given the significant effects of technological change per se on economic growth, a better understanding of the reasons behind the cyclical evolution of output and technology is important from a policy perspective. *Fifth*, our model elaborates on the role of basic and applied R&D mechanisms in the growth process. It shows that the influence of these two R&D types on the long-run growth process is very different indeed. *Sixth*, the model shows that in a competitive equilibrium the allocation of researchers between basic and applied R&D will generally be inefficient. We will also show that this inefficient allocation of R&D may give rise to situations in which the introduction of a carbon tax may yield a double dividend.

The latter findings are in line with Smulders and de Nooij (2002), who argue that if the allocation of R&D is inefficient, energy policy should be combined with a technology policy addressing the inefficiency in the R&D sector. We refine this analysis by distinguishing not only between two directions of technical progress (carbon-based versus non-carbon-based) but also between basic R&D and applied R&D. But in addition to this, our simulations also show that the effect of any policy scheme depends crucially on the existing technology structure causing policymaking to become highly path-dependent. If we are in a situation where carbon-based technologies form the dominant paradigm, then a carbon tax will – as a side-effect – move R&D resources away from applied R&D on the existing technologies towards basic R&D on new technologies, and this side effect will reduce the market failure in the R&D sector. Thus the carbon tax alone tends to reduce carbon-based fuel consumption and increase growth. Tax recycling as a subsidy on non-carbon-based fuels will reinforce the carbon reduction, and tax recycling as a subsidy on basic R&D will reinforce the growth effect, and depending on preferences, any of these schemes could be preferable.

The organization of the rest of this chapter is as follows. In section 2 we provide a short summary of the most important features of the STAGEPOST model, and our energy extensions of that model. Sections 3 and 4 show the results of some illustrative simulation exercises we have performed with the extended STAGEPOST model. In section 3 we describe the base run, and in section 4 we present some fiscal policy experiments. Section 5 contains some concluding remarks.

2. The STAGEPOST Model in Outline

2.1. GENERAL OVERVIEW

The organization of the extended STAGEPOST model resembles that of the Romer (1990) model. The extended STAGEPOST model consists of a perfectly competitive final output sector that combines labour and effective capital to produce output. Effective capital is an aggregate of individual GPTs that in turn consist of intermediates supplied to the final output sector under imperfectly competitive conditions. Intermediates are built according to the specifications from the blueprints obtained from the R&D sector that uses the market-value of these blueprints to pay for the resources used in producing them. The resources under consideration are R&D labour and R&D entrepreneurship.

The most important problem addressed in the STAGEPOST model is the "spontaneous" arrival of new GPTs as the result of basic R&D, and the subsequent expansion of that GPT through applied R&D. Both types of R&D are profit-incentive driven, as is usual in new growth models a la Romer (1990) and Aghion and Howitt (1992, 1998a). In our GPT set-up we assume that the most important components are invented first, while additional components invented through applied R&D grow less and less "promising" from a profitability point of view. The latter gives rise to decreasing returns to variety within a GPT thereby increasing the relative attractiveness of finding the core of a new GPT rather than continuing the expansion of existing GPTs by finding still newer peripheral components.[6] Thus the expansion of a GPT will eventually lose momentum. But this loss in momentum also increases the incentive to find a new core. In addition to this, it may well be the case that the core of a new GPT is not promising at all. Then our set-up implies that the R&D sector will devote most of its resources trying to find a new core rather than producing peripheral inventions for a failing GPT.[7] Thus we get cyclical growth patterns, implied by inherent technological expansion limits, rather than by the reallocation of resources between the final output sector and the R&D sector as one usually finds in GPT models.

Growth in the STAGEPOST model then is caused by symmetric horizontal innovation which increases the number of GPTs and quasi-vertical[8] innovation within each GPT, which effectively breaks the symmetry between GPTs again. Neither innovation is so drastic that it drives out older GPTs completely, however they can be asymptotically drastic.

2.2. THE PRODUCTION STRUCTURE

2.2.1. *The final output sector*

With respect to the production structure, we assume that a representative firm in the final goods sector produces output by using labour and effective capital that consists of a set of GPTs combined in an Ethier function (Ethier 1982). Each GPT contributes symmetrically to the effective capital stock. Each GPT itself is a composite input made out of different components, i.e. its core and peripherals. These components contribute to the GPT in an asymmetric fashion. We have:

$$Y = L_y^{1-\alpha} \cdot K_e^{\alpha} \qquad 0 < \alpha < 1 \tag{1}$$

where Y represents final output, L_y is production labour, and K_e is the effective capital stock, that is itself defined in terms of individual GPTs as:

$$K_e = \left(\sum_{j=1}^{A} z_j{}^{\alpha}\right)^{1/\alpha} \tag{2}$$

where A is the number of GPTs currently active, and j indexes those active GPTs. z_j represents the "volume" of GPT j. Equation (2) is a CES function with a symmetric contribution of all GPTs z_j to the level of effective capital K_e.[9] An obvious candidate to describe the inner structure of z_j is again a CES function:

$$z_j = \left(\sum_{i=0}^{A_j} c_{i,j} \cdot x_{i,j}{}^{\beta_j}\right)^{1/\beta_j} \tag{3}$$

where $x_{i,j}$ for $i > 0$ represents the i-th peripheral of GPT j and $x_{0,j}$ represents the size of the core of GPT j.[10] $c_{i,j}$ are the standard distribution parameters one normally uses with a CES function, while $1/(1-\beta_j)$ is the elasticity of substitution between all the components of GPT j. A_j is the number of peripherals belonging to GPT j.

In the very early stages of development of a GPT, A_j can actually be equal to zero, and in fact, even in the medium and long run, A_j can remain equal to zero, thus underlining the *ex post* character of what we consider to be real GPTs, i.e. technologies with a large number of peripherals. GPTs with a small number of peripherals A_j, will further be called "failed GPTs".

In order to simplify matters as much as possible, we make the following assumptions:

$$\beta_j = \alpha \;\forall\; j \tag{4.A}$$

$$c_{i,j} = c_{0,j} \cdot \left(\varsigma_j\right)^i \quad 0 < \varsigma_j < 1 \;\forall\; j \wedge i \geqslant 0 \tag{4.B}$$

Equation (4.A) effectively reduces the three-level organization of the production process to a two-level production function with asymmetric contributions of all components of all GPTs to final output, while Equation (4.B) puts a technically useful structure on the distribution coefficients of the implied aggregate production function. This structure allows us to write the production function in terms of an aggregate of functions of the cores of the various GPTs only, so from a practical point of view we can essentially forget about individual peripherals. This is a big technical bonus since we have to deal with a potentially large number of "real GPTs" simultaneously, and therefore potentially very many individual peripherals. The interpretation of Equation (4.B) is that in developing new components of a GPT researchers roughly know what to look for and where to look for it. Hence, they invent the core and then the most productive peripherals first, while subsequent peripherals contribute less and less to the productivity of the GPT as a whole.

Final output producers maximize profits under perfect competition conditions, conditional on the production structure as given by Equations (1)–(4.B), giving rise to the inverse demand functions for all the inputs in final output production.

2.2.2. *The intermediate goods sector*

The demand for GPT components is derived very much as in Romer (1990), i.e. intermediate goods producers face imperfect competition, as individual intermediate goods (or GPT components) are imperfect substitutes by assumption. They maximize profits conditional on the inverse demand function for their product arising from profit maximization in the final output sector:

$$x_{i,j} = L_y \cdot \left(\frac{p_{i,j}}{c_{i,j} \cdot \alpha} \right)^{-1/(1-\alpha)} \tag{5}$$

In the STAGEPOST model, it is assumed that the components can be produced using raw capital only, where each component i of GPT j takes η_j units of raw capital $k_{i,j}$ to create one unit of the component $x_{i,j}$, the marginal production costs are equal to $\eta_j \cdot r$, where r is the interest rate and where we have ignored the depreciation of capital. Because each component has its own market niche (as described by Equation (5)), the profit maximizing rental price of each component is easily obtained as:

$$p_{i,j} = mc_{i,j}/\alpha = r \cdot \eta_j/\alpha \tag{6}$$

which is the familiar Amoroso Robinson condition for profit maximization under imperfect competition, and where $mc_{i,j}$ is the marginal cost of component i of GPT j. Using Equation (6) to obtain the profit flow $\pi_{i,j}$per component i of GPT j, we find:

$$\pi_{i,j} = L_y \cdot c_{0,j}{}^{1/(1-\alpha)} \cdot \varsigma_j^{i/(1-\alpha)} \cdot (1-\alpha) \cdot \alpha^{(1+\alpha)/(1-\alpha)} \cdot \left(r \cdot \eta_j\right)^{-\alpha/(1-\alpha)} \tag{7}$$

Equation (7) has several interesting features. First, profits of a component rise with the overall level of final output production as proxied by L_y. Second, they fall with a rise in the production cost of a component (i.e. $r \cdot \eta_j$). Third, they fall with the peripheral index i (since $0 < \varsigma_j < 1$). The latter is one of the most important drivers of the overall behaviour of the model. Note, moreover, that for constant values of L_y and r, *ex post* profit flows are constant too. Under these assumptions, the present value of the profit stream associated with using component i of GPT j, i.e. $PV\pi_{i,j}$, would be given by:

$$PV\pi_{i,j} = \pi_{i,j}/r \tag{8}$$

As in Romer (1990) and Aghion and Howitt (1992), we assume that these profit flows in the intermediate sector are captured by the respective R&D sectors that created the designs for all individual components.

2.2.3. *The R&D sector*

We assume that each innovation, whether basic or applied, is the result of innovative activities in the R&D sector. Which type of R&D is done depends on the profitability of adding peripherals to an already existing technology (for which the core already exists) versus creating a completely new technology (for which a new core is required). As is clear from Equation (7), the profitability of a peripheral falls the later it is introduced, so that pursuing basic R&D becomes increasingly more profitable than doing applied R&D, and research labour gradually moves from applied research into basic research again. This gradual movement is due to our assumption that marginal R&D productivity is decreasing for both types of activities.

Our motivation for considering different R&D processes for peripherals and cores is our perception that the invention of a technology core requires "something more fundamental" than the further development of a technology by adding peripherals. We capture this difference by differentiating their contribution to total production. However, we also feel that finding a (core of a) potential GPT is subject to more uncertainty than finding a peripheral once a new technological "proto-paradigm" has arrived in the form of a core of a potential GPT. We model this by assuming that the R&D process itself is uncertain, first, because it is not possible to predict with complete certainty the exact arrival time of the core of a new GPT, and second, because it is not possible to predict with complete certainty the actual characteristics of a potential GPT. These characteristics are the inherent productivity of the next GPT (i.e. $c_{0,j}$), the user costs of the peripherals (i.e. η_j), the scope for extension (i.e. ζ_j) of the core, associated applied research productivity (i.e. δ_j, cf. Equation (9.B)), and research opportunities (i.e. μ_j, cf. Equation (9.A)). In order to model the uncertainty associated with basic research, we assign random values drawn from a uniform distribution to the characteristics of each GPT, which are unknown until the core of that GPT is actually invented. Applied R&D can only begin after the core has been introduced. At this point in time the GPT characteristics become publicly known, and for reasons of simplicity, we assume that there is no uncertainty regarding the characteristics of new peripherals added to the GPT through applied R&D.[11] Of course, there is still a random element in applied R&D because the actual arrival of an applied R&D invention depends on random draws from a Poisson distribution.

Even though we explicitly distinguish between basic and applied R&D, we do model both types along very similar lines. In both cases R&D gives rise to innovations that arrive according to a Poisson probability distribution. As in Aghion and Howitt (1992), we assume that the level of R&D activity directly and positively influences the arrival rate of innovations. But unlike Aghion and Howitt, we assume that there are decreasing marginal returns to current R&D, giving rise to an effective arrival rate of innovations given by:

$$\lambda_j = \mu_j \cdot R_j^e \tag{9.A}$$

$$R_j^e = \delta_j \cdot R_j^\beta \tag{9.B}$$

where λ_j is the arrival rate of innovations associated with R&D process j (process 0 is associated with the basic R&D necessary to find the core of the next GPT, whereas $j > 0$ represent the processes necessary to find the peripherals of GPT j). μ_j is the arrival rate at a unit level of effective R&D, R_j^e. Equation (9.B) shows how effective R&D uses R&D labour R_j with a "productivity" parameter equal to δ_j, and where $0 < \beta < 1$ ensures that the marginal product of R&D labour falls with the level of R&D activity. The latter is another main feature of the model, since it ensures that in combination with Equation (8), the marginal benefits of doing applied R&D and basic R&D are asymptotically falling to zero for increasing levels of R&D activity. The marginal benefits rise to infinity for levels of R&D activity that fall asymptotically to zero. The latter ensures that there will always be an incentive to employ a non-zero volume of R&D workers on any project, however bleak the prospects for success may be.

Expected profits arising from R&D activities are maximized by hiring additional R&D workers at the ruling R&D wage, until the value marginal product of doing R&D matches the marginal cost of hiring an additional R&D worker. The expected marginal benefits from doing R&D on the i-th peripheral of GPT j are defined in terms of the (expected present value of the) profit flows arising from the use of a (new) component in the final output sector, and hence the incentive to produce that component in the intermediate goods sector. The expected benefits from doing R&D would therefore be the expected present value of all the profits in the intermediate goods sector that are directly associated with the production of a particular component. Hence, the marginal benefits of doing R&D would then be given by:

$$MB_{i,j} = \frac{\partial(PV\pi_{i,j} \cdot \lambda_j)}{\partial R_j} = \frac{\partial(PV\pi_{i,j} \cdot \mu_j \cdot \delta_j \cdot R_j^\beta)}{\partial R_j} \tag{10}$$

In Equation (10), $PV\pi_{i,j} \cdot \lambda_j$ is the *expected* present value of the allocation of R_j R&D workers.[12] As the free mobility of R&D labour between its various uses implies that the marginal benefits for different R&D activities should be the same, we obtain for the optimum ratio of applied R&D and basic R&D workers:

$$\frac{R_j}{R_0} = \left(\frac{PV\pi_{A_j+1,j} \cdot \mu_j \cdot \delta_j}{PV\pi_{0,A+1} \cdot \mu_0 \cdot \delta_0}\right)^{1/(1-\beta)} \equiv \varphi_j \tag{11}$$

where A_j is the number of peripherals of GPT j, and A is the total number of active GPTs. Equation (11) shows that higher (expected) profit flows on a peripheral in some GPT will divert R&D resources into further expanding that GPT. Such an expansion is promoted by an R&D process that is relatively efficient (high δ_j), or a process where innovations are relatively easy because of ample "fishing" opportunities (high value of μ_j). Because total R&D uses exhaust the available and exogenous supply of R&D workers R, we readily find:

$$R_0 = R/(1 + \sum_{j=1}^{A} \varphi_j) \tag{12.A}$$

$$R_j = \varphi_j \cdot R_0 \tag{12.B}$$

A fall in the present value of the expected profit flow associated with peripheral j will therefore decrease the corresponding φ_j and for given R, the corresponding value of R_j will fall, thus leading to a decrease in the rate of expansion of existing GPTs and an acceleration of the arrival rate of new GPTs.

Finally, it should be noted that as R&D workers are paid their value marginal product, and as there are decreasing returns to R&D, there is also a profit flow that should be regarded as the reward for R&D entrepreneurship.

2.3. LOVE OF VARIETY AGAIN

The organization of capital in the form of different GPTs with differential impacts on effective capital ultimately results in the growth of capital productivity, hence of output itself, for a given amount of capital and labour. This is easy to see, since after some manipulation of the relation between effective capital and the size of the core of some GPT j, as well as the capital costs of building the core (and the corresponding peripherals), we get:

$$z_j = K_j \cdot c_{0,j}{}^{1/\alpha} \cdot \left\{(1 - \varsigma_j^{(1+A_j)/(1-\alpha)})/(1 - \varsigma_j^{1/(1-\alpha)})\right\}^{1/\alpha - 1} /\eta_j \tag{13}$$

The "GPT" productivity of "raw" capital K_j, i.e. z_j/K_j, depends positively on the size of the contribution of the core (i.e. $c_{0,j}$), which we have already referred to as the intrinsic productivity of the GPT, positively on the scope for extension of the GPT (i.e. ς_j), negatively on the unit raw capital cost (η_j), but most importantly, it depends positively on the number of peripherals of the GPT, (i.e. A_j). The latter is the Love of Variety effect implied by the concavity of the GPT in its individual components (cf. Equation (3)). Love of Variety works at two different levels in this model, therefore, as opposed to Romer (1990). It works at the component level within each individual GPT, but also at the level of the GPTs that together constitute effective capital.

2.4. ADDING THE ENERGY DIMENSION

We introduce energy into the STAGEPOST model by means of the following modifications. First, we distinguish between two types of GPTs that use either carbon-based or non-carbon-based fuels. Each type of GPT results from directed search for a core that is either carbon-based or non-carbon based. Hence, we now have **two** basic R&D processes that are active at any moment in time as well as a potentially large number of applied R&D processes on each carbon-based and non-carbon-based GPT. Second, the marginal costs of the components of these GPTs now also include carbon- and non-carbon-based fuel costs, next to capital cost. In fact, we assume that fuel–capital ratios at the component level are fixed.[13] However, these fuel–capital ratios may differ between GPTs as this is another random GPT characteristic.

For environmental quality we adopt the simplest thinkable formulation:

$$Q_t = Q_t^{\max} - F_t^C \tag{14}$$

where $Q^{\max}$ is the maximum attainable quality, which can only be reached if no carbon-based fuels are used. F^C is the total amount of carbon-based fuel that is being used in production. To simplify matters even more, we assume that pollution does not accumulate. This means that in the model, if carbon-based fuel consumption would be abolished altogether, the environment would return to its maximum quality immediately, which is obviously not a realistic assumption. For ease of exposition, we assume that the consumption of non-carbon-based fuels does not harm the environment at all.

2.5. MODEL CLOSURE

The model is closed by assuming that fuel prices are exogenously determined. We also assume that the interest rate is exogenously fixed. Usually, a fixed interest rate is "defended" by means of a small country assumption. However, the latter assumption would be hard to reconcile with technological change all being generated within a small country, rather than largely being imported from abroad. It should also be noted, however, that in Love of Variety growth models using a production structure similar to ours, the expansion of the number of varieties effectively works at the aggregate level as labour augmenting technical change in the context of the Solow growth model, thus giving rise to a constant steady state real interest rate. Hence, the *assumption* of a constant real interest rate, while admittedly not very elegant, is at least not inconsistent with the steady state properties of a Love of Variety growth model. We have had to make this assumption because of numerical difficulties we encountered trying to endogenize the calculation of the real interest rate that equalizes the demand and supply of capital at a given saving rate. The reason for these convergence problems is that the actual arrival of an innovation leads to corresponding discontinuities (i.e. "jumps") in the model variables that sometimes made it impossible to find a simultaneous numerical solution to the model. This holds *a fortiori* for periods in which R&D activity is relatively successful, or where the composition of R&D is changing rapidly, i.e. exactly the periods we are interested in *a priori*.

Given the assumptions above, the consolidation of all income and expenditure streams in this model-economy, gives rise to real disposable income of households that consists of the value of final output, less capital and fuel costs. Capital and fuel costs are the only "cost" items (including profits in the intermediate goods sector and the R&D sector as balancing items), that do not function as an income stream into another domestic sector at the same time, including households. Again, this is a practical short cut we take, as ideally we would have to specify how the capital market would finance both capital accumulation activities and risky R&D activities, especially those still having to result in a new component, as in Wälde (1999, 2004), for example. For our present illustrative purposes this is a bridge too far, and we have to leave the inclusion of explicit investment decisions of this kind in a downsized analytical version of our model till later.

3. Base Run Simulation Results

We illustrate the working of the model by means of a simulation analysis. Since some of the GPT characteristics are relatively abstract concepts, it would be very difficult to find real-world data that could be used directly for calibration purposes. Instead, we choose a set of arbitrary parameter values. This approach allows us to draw only qualitative conclusions for policy analysis.

In order to illustrate the behaviour of the model, we first present the results of a base run. Figure 1 shows how the number of available GPTs rises over time. We assume that the economy starts with just one carbon-based and one non-carbon-based GPT. Figure 2 shows the number of peripherals that are developed for each GPT. Interestingly, some GPTs develop only very few peripherals or none at all. These are "failed" GPTs, consisting of only a core with few or no peripherals. But there are also "true" GPTs such as C03 or N04, which develop dozens of peripherals.[14] We see these "failed" GPTs and "true" GPTs in our *ex post* perspective, but the researchers who developed them were seeing them as potential GPTs, because they could only guess at their true characteristics *ex ante*.

The number of peripherals allows us to say something about the usefulness of a potential GPT, but it does not tell us very much about the economic impact

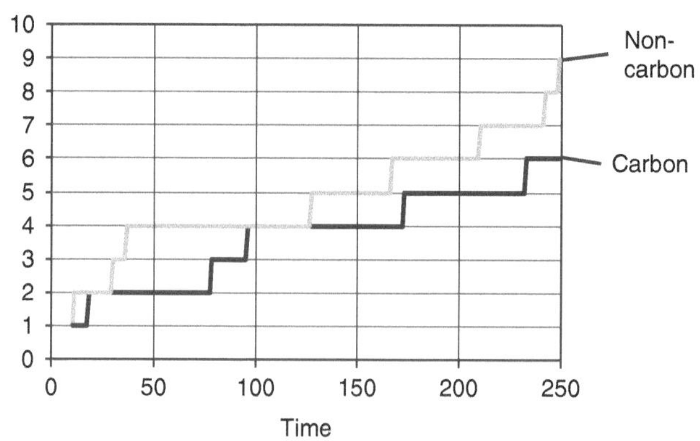

Figure 1. Number of GPTs.

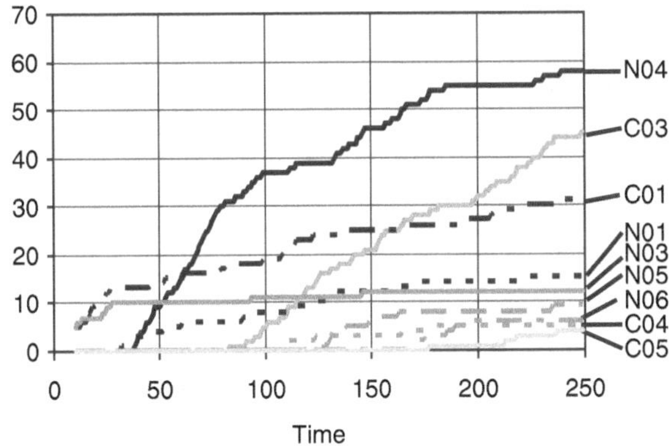

Figure 2. Number of peripherals by GPT.

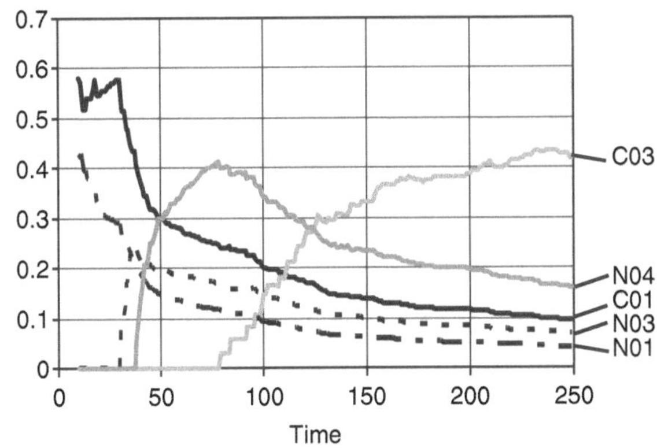

Figure 3. Relative contribution to K_e by GPT.

of the GPT. However, one important general feature of a GPT is that it affects a large share of the economy. In order to measure this feature we show the Eulerian contribution of each GPT to effective capital K_e in Figure 3 as a measure of the relative contribution to output (through the effective capital stock) of each individual GPT.[15] The curves in Figure 3 show the development over time of the share of a specific technology in effective capital. They are not diffusion curves in the usual sense, even though they look like it. For, diffusion takes place as entrepreneurs adopt the same technology at different points in time because these entrepreneurs are somehow intrinsically different. Here we have a different situation, in that the core idea gradually diffuses into the economy as more and more peripherals are invented as applications/extensions of the core invention.

The curves in Figure 3 show a cyclical pattern, which is consistent with some long-wave views on economic development. Freeman and Perez (1988), for example, identify five Kondratieffs in economic history since the late 18th century. Such Kondratieffs are characterized by the dominant GPT of their times, for instance the "steam power and railway Kondratieff" in the mid-19th century or the "information and communication Kondratieff" that started in the late 20th century. Note that in contrast to many long-wave theories, there is nothing mechanistic in the coming and going of Kondratieffs in our model. Due to the structure of the model every GPT will at some time run out of further extension possibilities, and the search for a new GPT begins, but the length of these "long waves" is endogenously determined and highly variable. In the current simulation run, for example, the first Kondratieff is dominated by the initial technologies C01 and N01. It is quickly succeeded by another Kondratieff which is dominated by N04. Around the year 120, N04 is succeeded by C03, which dominates the next Kondratieff lasting until the end of the simulation period. Other simulation runs show a similar picture. Usually, the length of a Kondratieff is in between 10 and 40 years, but sometimes an extremely successful GPT is invented which remains dominant for 100 years or more.

Figure 4 shows how the dominance of a certain technology determines the economy's fuel mix. During the N03 Kondratieff, non-carbon-based fuel consumption is much higher than carbon-based fuel consumption. During the transition to C03, however, non-carbon-based fuel consumption levels off, while carbon-based fuel consumption quickly rises. At the end of the simulation period, the economy consumes a balanced mix of both fuels. There is no long-run tendency towards either fuel, because we assume that carbon-based fuel technologies are intrinsically just as productive as non-carbon-based fuel technologies. Thus, the R&D sector has no reason to concentrate on either fuel, and over the long run the economy can be expected to develop just as many carbon-based GPTs as non-carbon-based GPTs.

Figure 5 shows the allocation of R&D workers between basic and applied R&D. We have chosen the total number of researchers – plotted on the vertical axis – to be equal to five. Just as in Figure 3, we clearly observe cycles in R&D. Whenever an attractive GPT is invented, researchers move into applied R&D on the new GPT. In year 31, for instance, the freshly invented N03 absorbs almost

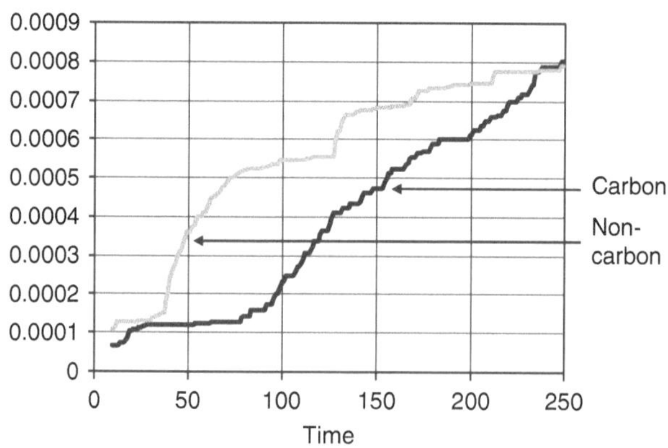

Figure 4. Fuel consumption by type of GPT.

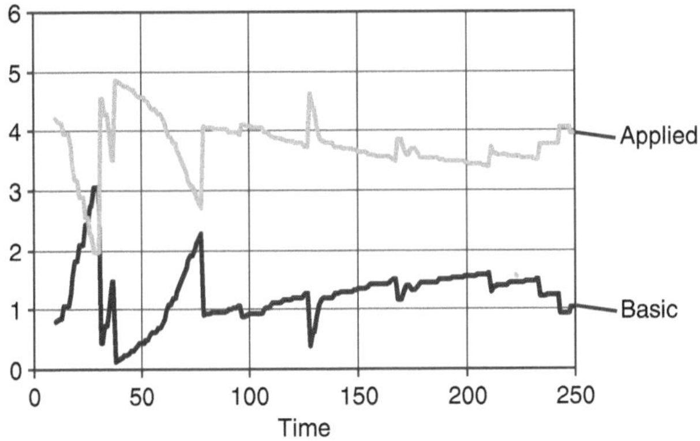

Figure 5. Basic vs. applied R&D.

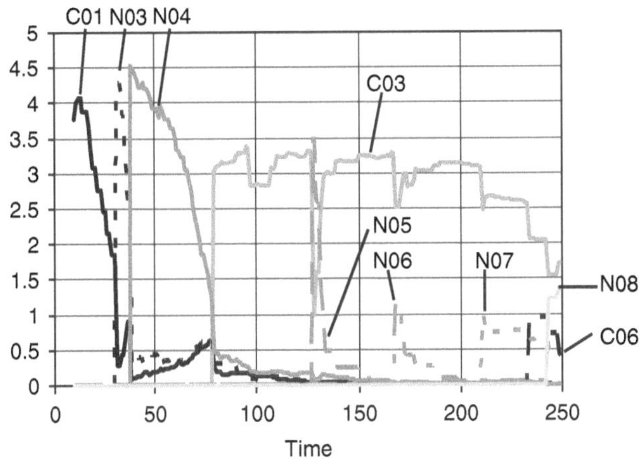

Figure 6. Applied R&D by GPT.

all of the R&D workers, who are busily developing peripheral applications based on that GPT. As the extension possibilities of N03 are being exploited, further R&D on N03 becomes less attractive, and researchers gradually return to doing basic R&D.

Figure 6 shows the amount of applied R&D on each GPT. Only successful GPTs are shown for the sake of visual clarity. We can see how researchers move away from applied R&D on N04 as that GPT runs out of extension possibilities. After the introduction of a new GPT, there is a "jump" in the R&D sector since there is a massive and instantaneous movement from basic R&D into applied R&D.[16] The figure can help to explain the continued dominance of C03. It is a GPT with a very large scope for extension, in model terms its ζ is very close to one. New technologies, such as N05 or N06, attract a certain amount of applied R&D for a short time, but as these technologies offer a limited scope for extension, R&D workers move back into applied R&D on C03.

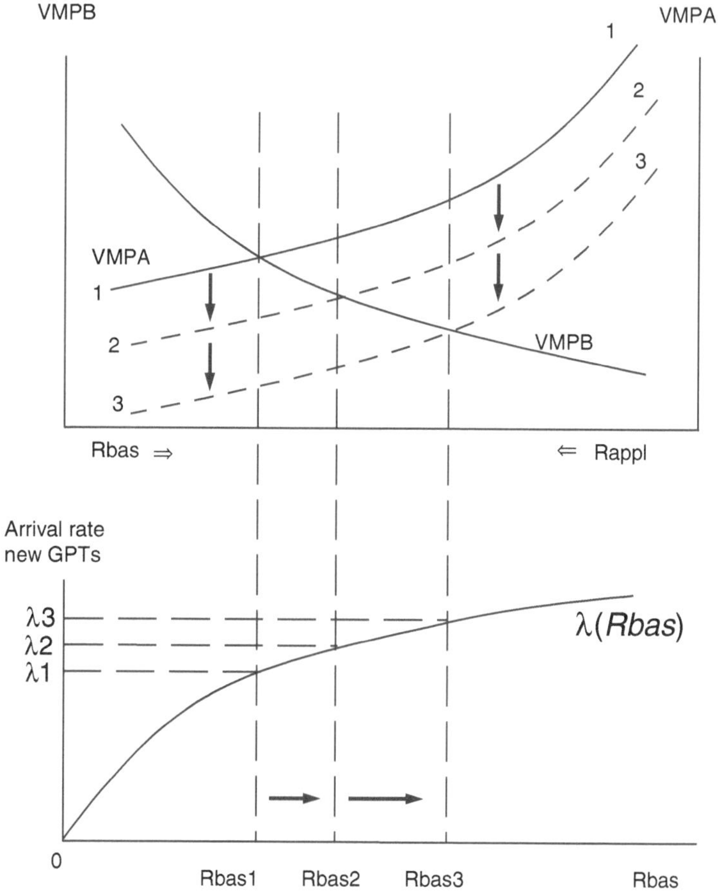

Figure 7. Allocation of R&D.

The economic intuition behind these R&D movements can be shown graphically in Figure 7. The horizontal axis represents the (given) total number of R&D workers, who are either working in basic R&D or applied R&D. Moving to the right, the ratio of basic R&D to applied R&D increases, and the other way around. The two curves depict the value marginal product (VMP) of employing basic and applied R&D.[17] Since we have assumed diminishing returns ($\beta < 1$), the curves are downward sloping The point of intersection of both VMP curves determines the profit maximizing allocation of researchers, as both profits (and R&D wages if R&D workers are paid their value marginal product, as we assume) cannot be increased by moving from the low VMP R&D activity into the high VMP R&D activity.

If, in this setting, applied R&D is successful, a new peripheral is developed, and we know that the next peripheral will be less productive (because $\zeta < 1$). Thus, the VMP curve associated with applied R&D shifts downward. At the new point of intersection, more researchers are working in basic R&D than before.

The bottom half of the figure shows the arrival rate of new GPTs as a function of the number of active basic R&D workers. We can see that as more and more researchers move into basic R&D, the arrival rate increases, and the arrival of the next GPT becomes increasingly more likely. When the next GPT with a lot of development potential actually arrives, new possibilities for applied R&D arise. The VMP curve of applied R&D shifts up (cf. Equations (7), (8), and (10)), and the allocation of R&D workers changes again in favour of applied R&D. Thus, the alternating arrivals of GPT cores and peripherals generate cycles in the R&D sector, with researchers moving back and forth between the two R&D activities, thus generating growth.

In order to accommodate the two different basic R&D processes aimed at finding the carbon- and non-carbon-based "cores" that were referred to earlier, Figure 7 can be extended as follows. In Figure 8, the left-hand panel is a copy of the top panel of Figure 7. However, the left panel of Figure 8 now refers to carbon-based GPTs as indicated by the superscript "C". Similarly, the right panel of Figure 8 refers to non-carbon based GPTs, as indicated by the superscript "NC". The arrival of a new non-carbon core, for example, now raises the value marginal product of applied R&D workers in the non-carbon R&D sector both relative to basic R&D in that sector, and to all R&D in the carbon R&D sector. Hence the non-carbon R&D sector will experience an inflow from the carbon-sector, whereas in both R&D sectors basic R&D will drop in absolute terms. The latter also goes for applied R&D in the carbon sector. In Figure 8, the initial allocation of R&D workers between sectors and between basic and applied R&D within both sectors

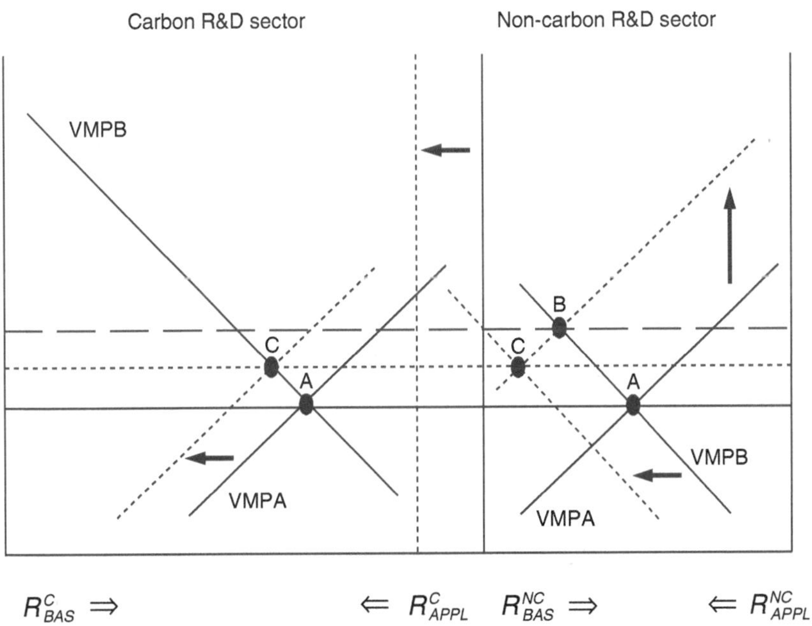

Figure 8. Allocation of R&D, Carbon versus Non-Carbon.

is depicted by the solid lines. The final equilibrium allocation of R&D workers after the arrival of the core of a non-carbon based GPT is depicted using the dotted lines. As in Figure 7, this arrival would lead to an upward shift of the VMPA curve in the non-carbon R&D sector. For a given volume of non-carbon R&D workers, this would raise the wages of the latter from point A to point B, and change the allocation of R&D workers within the non-carbon sector between basic and applied R&D, accordingly. The resulting wage differential between the carbon R&D sector and the non-carbon R&D sector now invokes an inflow of R&D workers into the non-carbon sector, and therefore leads to an expansion of non-carbon-based R&D and a fall in the level of carbon-based R&D, as depicted by the left-shift of the vertical in the middle of Figure 8. This inflow would also reduce the upward pressure on R&D wages in the non-carbon sector somewhat, leading to a movement from point B to point C as the final equilibrium R&D constellation in both sectors.[18]

The arrival of successful GPTs generates cycles in the growth rate of the economy as we can see in Figure 9, which shows the growth rate of real disposable income (RDI). Successful GPTs have two effects on growth. First, they raise output immediately in the period after their introduction, simply because of our Love of Variety structure. But second, and more important, they can raise the average growth rate over a period of several decades, as researchers are exploiting the new possibilities for applied R&D.

The actual growth impact of new GPTs, of course, depends on the intrinsic characteristics of the GPT, which in turn determine its pervasiveness, given the "general GPT environment".[19] In the extreme case, i.e. a total failure with no peripherals at all, we have only a short-lived growth hike due to the Love of Variety effect. This uncertainty about future growth rates is also a realistic feature of the model: we know with almost complete certainty that a new successful GPT will arrive and that it will set off a period of high growth, but we can never know exactly when this will happen or whether it will actually be the next GPT that turns out to be a big success.

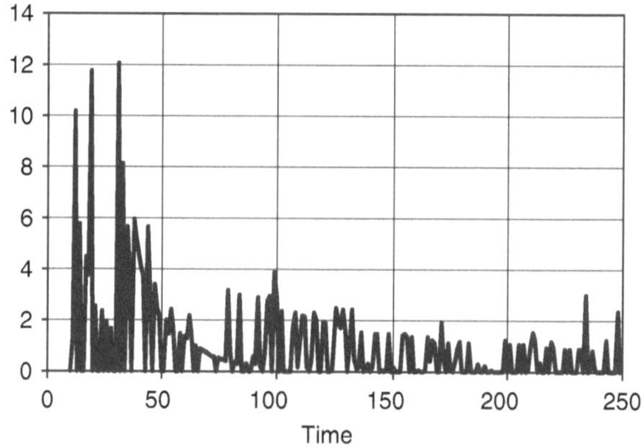

Figure 9. Real disposable income growth (%).

As regards the development of welfare, we have experimented with a CES function incorporating environmental quality and real disposable income as arguments. Generally speaking, we have found that as consumption and environmental quality become better substitutes, output growth also results in welfare growth, as increased consumption possibilities can more than make up for decreased environmental quality. In the extreme case where consumption and environmental quality would be perfect substitutes, it is always possible to make up the utility loss from environmental degradation through higher consumption. For low values of the elasticity of substitution between consumption and environmental quality, we find that the effect of environmental degradation may be so large that utility actually starts falling at some point in time: because through growth the marginal utility of consumption diminishes vis-à-vis that of environmental quality, it becomes too difficult in the end to raise welfare through "dirty" growth.

4. Fiscal Policy Experiments

4.1. INTRODUCTION

In most Western countries, governments are concerned about carbon-based fuel consumption for at least two reasons. First, carbon-based fuels lead to unavoidable CO_2 emissions, which are damaging the environment. Improvements in energy efficiency and CO_2 sequestration may ameliorate this problem, but they cannot solve it in the very long run. Second, most oil deposits are situated in countries that are politically unstable. Therefore, almost all Western governments are implementing policies to reduce their dependence on carbon-based fuels.

However, it is not so much the *volume* of fuel consumption that worries governments and environmentalists; it is primarily the *mix* of fuels. If it were possible to substitute non-carbon-based fuels (e.g. hydrogen produced from non-carbon-based energy sources) for carbon-based fuels (e.g. oil and gasoline) at a large scale, the problem could be solved. One would simply have to discourage carbon-based fuel consumption by means of a carbon tax. The question then is what to do with the tax revenue. Some have proposed to use the tax revenue to subsidize the consumption of non-carbon-based fuels. This would alter the relative price of non-carbon-based fuels, and the resulting substitution effect might lead to the desired change in fuel consumption. Alternatively, one could subsidize R&D in the hope that new technologies based on non-carbon-based fuels are developed.

We now examine the impacts of two different energy policy schemes on growth and the environment (and ultimately welfare).

4.2. POLICY 1: A CARBON TAX AND NON-CARBON-BASED FUEL COST SUBSIDY

Our first policy experiment is going to be a proportional tax on the consumption of carbon-based fuels, which is to be recycled in the form of a subsidy

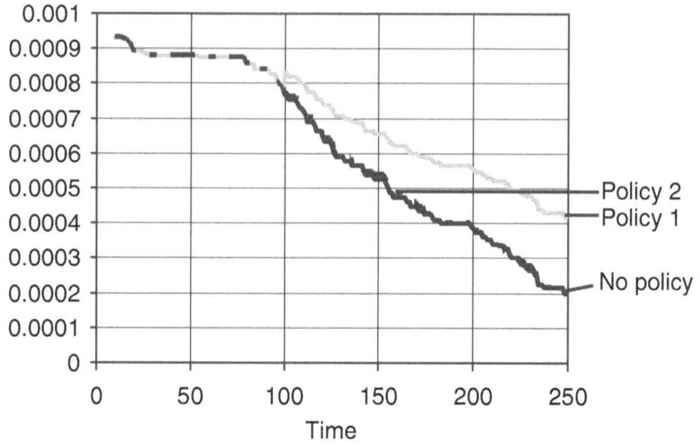

Figure 10. Environmental quality.

on non-carbon-based fuel consumption. This is going to generate a substitution effect away from carbon-based fuels and towards non-carbon-based fuels. We introduce the policy in the year 100.[20] Figure 10 shows that the policy is effective in achieving the goal of increasing environmental quality. The introduction of the tax/subsidy leads to an immediate fall in carbon-based fuel consumption, and due to the one-to-one relationship, environmental quality is immediately improved. As the economy grows, carbon-based fuel consumption grows along, but it remains far below its base run value, and although environmental quality keeps falling, it remains way above the baseline value.

Of course, the government will not only be concerned about carbon-based fuel consumption, it will also be concerned about the policy's effects on output. The data show that the policy leads to an initial drop in RDI. The initial RDI loss is not surprising because the tax/subsidy creates a distortion in the final output sector. However, this RDI loss is eventually overcome: In the year 166 RDI under policy 1 surpasses the base run value. From that year on, RDI is actually higher than in the base run. Under these circumstances, the government may be able to realize a double dividend in this scenario. Compared to the base run, the policy increases output in the long run and reduces carbon-based fuel consumption immediately. One would have to weigh the present value of the short run RDI loss against the present value of the long run RDI gain, and the outcome depends on a subjective discount rate.

Figure 12 shows why there is an increase in RDI in the long run. The number of GPTs grows faster under policy 1, so that at the end of the simulation period there are now 16 GPTs instead of 14. The reason for the faster arrival of GPTs is that although the policy targets only the fuel market, it also has an effect on the allocation of R&D resources. By lowering the relative price of non-carbon energy, the policy raises the value of a patent on a non-carbon technology. The present value of these patents rises relative to that of the carbon-technology patents. This means, first of all, that there will be more applied R&D on non-carbon GPTs and less on carbon GPTs. Furthermore, since the dominant C04 offering the most

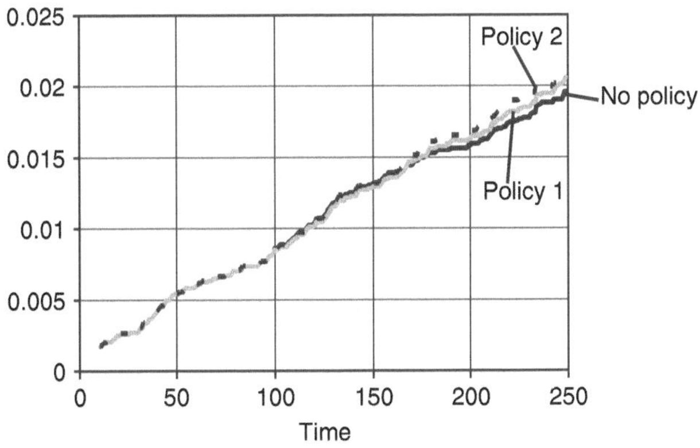

Figure 11. Real disposable income.

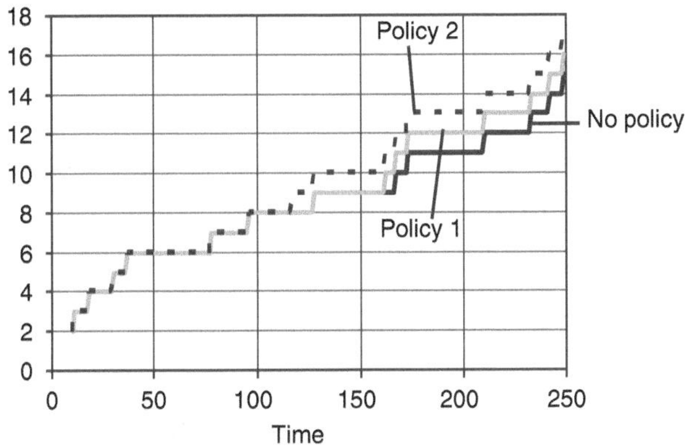

Figure 12. Number of GPTs.

applied research opportunities is carbon-based, there will be a larger incentive to do basic R&D on finding the next non-carbon-based GPT. Thus, there is a shift from applied R&D to basic R&D, and we have already shown that due to the externality of basic research, there is generally too little basic research in equilibrium. Therefore, the policy-induced shift towards basic R&D leads to a higher long run growth rate.

4.3. POLICY 2: A CARBON TAX AND BASIC R&D SUBSIDY

Realizing that the increase in growth was only possible because the market failure in the R&D sector was partially overcome, one might be tempted to tackle the problem at its root by subsidizing basic R&D directly. Therefore, the next policy experiment is the same carbon tax as before, with the difference that we now use the tax revenue to finance a subsidy on basic R&D.

Figure 11 shows that this policy is indeed very successful in terms of growth. There is still the initial RDI loss, but it is overcome in the year 118, much sooner than under policy 1. Thus, in a net present value analysis, this policy has a much better chance of yielding a positive result than policy 1 did. The reason for this increase in growth is that the reallocation of R&D activity is much larger, and so the arrival rate of new GPTs is sped up considerably. Under policy 2, C05 is invented in the year 117, eight years earlier than under policy 1. Although this technology is a "failed" GPT, it still contributes to RDI, and due to its earlier invention, RDI is higher under policy 2.

Although policy 1 is highly beneficial for growth, it is less effective in reducing carbon emissions, for two reasons: First, the change in the relative price of non-carbon-based fuel is smaller. Under policy 1 the price effect of the tax on carbon was reinforced by the subsidy on non-carbon. Now there is only the tax effect, because the subsidy on non-carbon-based fuel is replaced by another subsidy. With a smaller relative price change, the substitution effect is weaker. The second reason is that the increase in growth again raises the demand for both carbon- and non-carbon-based fuels. As a result, from the year 174 on, carbon-based fuel consumption is actually higher than in the base run, and environmental quality is lower than in the base run.

4.4. POTENTIAL WELFARE EFFECTS

The potential welfare effects of any policy experiment depend crucially on the assumptions we make about the utility function, especially with respect to the elasticity of substitution between consumption and environmental quality. However, in the long run both policy schemes are bound to be welfare-improving. In the case of policy 1, this positive effect comes mainly from the cleaner environment, but also from the increase in RDI. In the case of policy 2, the positive welfare effect comes from higher RDI, which offsets the increase in pollution, the more so if environmental quality and consumption are good substitutes. Second, with both policy schemes there may be an initial negative impact on welfare if substitutability is high. This is intuitively plausible, because the initial RDI loss itself that is caused by the introduction of the tax, is welfare-decreasing. But over time, the policy schemes increase economic growth, thus raising welfare again.

4.5. ROBUSTNESS OF RESULTS

In order to get an impression of the robustness of the results we have run a number of different simulations. It turns out that policy 2, the subsidy on basic R&D, is always growth-increasing, because it provides a remedy to the market failure in the R&D sector. Of course, the effect of the policy also depends on the size of the tax/subsidy. If we set the subsidy on basic R&D extremely high, we will at some time reach the point where basic R&D is actually too high, and growth is slowed down again. There must be an optimal subsidy that exactly equates the social marginal benefits of basic and applied R&D. The exact value of this optimal subsidy is difficult to determine in a model of this complexity, however. The ultimate reason for this is that the arrival of new GPTs and the subsequent expansion of these GPTs is governed by random events, which would allow us to equate these marginal social benefits only in an average sense.

In addition to this, the model generates path dependencies which in combination with the randomness referred to above, makes it impossible to define generic policy prescriptions even in an average sense. To illustrate the point, in the scenario described above a carbon-based GPT happened to be the dominant one for most of the simulation period. Under these circumstances, policy 1 will induce researchers to move away from applied R&D into basic R&D, because they hope to discover a new GPT that is based on non-carbon-based fuels, and we have shown that reallocating resources towards basic R&D speeds up growth. If, on the other hand, a non-carbon GPT happens to be dominant, the results are exactly reversed: resources shift from basic R&D towards applied R&D on the non-carbon GPT, and consequently growth slows down. Therefore, the growth effect of policy 1 depends crucially on the existing technology structure.

To check our intuition behind the effects of policy 2, in which tax revenues are used to subsidize basic R&D, we have run simulations with still another policy scheme in which applied R&D is subsidized. This policy 3 is never the optimal one, because it will yield the same substitution effect as policy 2 (and a smaller one than policy 1), but it exacerbates the misallocation between basic and applied R&D, and its effect on growth will generally be negative.

The choice between policy 1 and 2 depends on the preferences that are assumed. We have underlined that if consumption and environmental quality are regarded as good substitutes, policy 2 is likely to be optimal because it speeds up growth, albeit at the cost of higher pollution. But if they are considered to be poor substitutes, policy 1 is likely to be preferable, because it is more effective at bringing down pollution.

The scenario we have chosen for the simulation analyses is one where a carbon-based GPT happens to be dominant. Therefore, the economy is highly polluted, and environmental quality is the "scarce" good. If, however, the economy was in a situation where either output is low or "clean" technologies dominate, the policies may have fundamentally different effects. It is important to note that the optimal policy depends on the existing technology structure. Therefore, there may not be a universally optimal policy. It might be optimal, for example, to employ policy 2 at low income levels to speed up growth, and then, at higher income levels, to move towards policy 1 in order to curb carbon-based fuel consumption. More simulation experiments may yield further insights into these matters.

Finally, it should be mentioned that the effects of policies in this model are usually not very long-lasting. With this statement we mean that if a policy is removed – even after a very long time period – the economy tends to jump back towards a development path very close to the one it would have followed without the policy. This phenomenon is due to the "putty" capital we have assumed, and that is a standard feature of many endogenous growth models.[21] If a policy measure changes the relative price of any fuel, the capital stock is completely free to adjust instantaneously. In reality, capital is often “clay *ex post*”. That is, if the price of one type of fuel rises, it is not possible to immediately retrofit all engines to use the other type of fuel. A future version of the model will incorporate some of these “clay” capital features, and we expect the effects of policies to be more long-lasting than in the current model, thus emphasizing the importance of the actual timing of policy measures in the face of limited substitution possibilities *ex post*.

5. Concluding Remarks

We have shown that a large number of GPTs can coexist in a model that is still not excessively complex. With the structure that we have assumed, it is possible to generate GPTs that are based on a highly successful core technology that is extended with a number of peripheral technologies. The model creates long-run growth with a Schumpeterian flavour, because the introduction of a new core technology (a “radical innovation” in Schumpeterian terms) is followed by a quick succession of peripheral technologies (“incremental innovations”) that give rise to alternating phases of fast and slow growth. There is nothing mechanistic behind these “long waves”, although the demise of a GPT is certain in the long run because the possibilities of extension are limited, and new GPTs are sure to arise. These features give rise to a model economy capturing many of the stylized facts that we observe in actual economies over the long run.

Assuming that different GPTs are based on different energy sources, it is clear that this succession of GPTs will also generate long waves in the consumption of different energy sources. We have constructed a model with two energy sources, where each GPT is based on either carbon-based fuels or non-carbon-based fuels. Policies that discourage the use of carbon-based fuels can have a large impact on the allocation of R&D resources in such a framework.

Using this model, we have examined the impact of different policy scenarios. A common element in all policies was a tax on the consumption of carbon-based fuels, but the tax recycling was different in each scenario. The results associated with each policy scenario differed considerably. Thus, when we discuss the usefulness of a carbon tax, it is very important to know how the tax is going to be recycled. In addition, we have seen that the effect of such a policy also depends on the existing technology structure. If the dominant GPT is carbon-based, a carbon tax will speed up the search for new GPTs, thereby stimulating growth. But if a non-carbon-based GPT dominates, the carbon tax might result in excessive applied R&D on that GPT, thus delaying the arrival of new GPTs and reducing growth.

We have shown that if the competitive allocation of R&D resources is inefficient, policy may increase the growth rate. In such a case, the carbon tax alone is a second-best solution and should be combined with a policy that addresses the inefficiency in R&D allocation. A similar argument has been made by Smulders and de Nooij (2002), who examined the effects of energy conservation policy in a model with two directions of technical progress. In our model, however, the inefficiency results from a non-optimal allocation of resources between basic and applied R&D. Therefore, if growth were the primary policy objective, it would be easiest to tax applied R&D and subsidize basic R&D, without creating any distortions on the energy market. If one wishes to promote a carbon tax on the grounds of R&D efficiency, one would have to show that R&D on carbon technologies is inefficiently high. In fact, our model provides potential support to such an argument.

Concerning the coming energy transition, it is clear that the hydrogen system is not even yet in its infancy. If it is going to be a core technology, forming the basis for a potentially wide range of peripheral innovations, a lot of basic research will have to be done before the system can actually be implemented. The traditional, oil-based energy system, by contrast, is a very mature system. No more basic research is required, only a few more peripheral applications will be feasible. Since basic research generates positive externalities it is presumably underprovided by the market, and a carbon tax combined with a subsidy on basic R&D would be a very effective tool to increase not only the speed of transition, but also the growth rate of the economy as a whole. A consumption subsidy on non-carbon-based fuels would not be the most effective solution, because it would end up supporting current producers of non-carbon-based fuels, and they are not necessarily the ones who develop new non-carbon technologies. The most effective way to speed up the transition towards a non-carbon energy system will be to promote basic R&D on non-carbon energy technologies.

The current model has been specified in a fairly minimalistic way, since it was only meant to serve the purpose of illustrating the principles involved. But it could be extended along several lines. One interesting extension would be to include labour mobility between the final output sector and the research sector. This would enable us to assess the impacts of policy measures on wage differentials or unemployment. It will also make our results more comparable to other GPT models, because in such models it is the movement of labour between the final output and the research sector that generates output booms and busts.

Furthermore, it has been argued that rising oil prices will contribute to the demise of the oil energy system and the rise of a hydrogen economy. We may also incorporate this effect into the model. For the time being, we have assumed constant real prices of capital and energy. We could allow energy prices to change exogenously over time, or include supply functions for carbon- and non-carbon-based fuels.

Another promising route of further investigation would be to look at the optimal timing of policies. We have argued that a growth-promoting policy might be optimal at low income levels, whereas a pollution-curbing policy might be optimal at higher income levels. These issues depend on preferences, specifically the elasticity of substitution between consumption and environmental quality. Especially with regard to the timing of policies, it would be interesting to incorporate clay capital and different GPT vintages into the model. With clay capital the *ex-post* substitution possibilities between vintages are zero, and the timely execution of policies becomes much more crucial than in the current model. In this context, and considering the randomness and path dependencies involved, Monte Carlo experiments may be helpful in finding robust policies that get the policy job done in almost all thinkable circumstances.

Notes

1. See, for example, the February 2005 issue of the *Scientific American*, or the August 2004 issue of *Physics Today*, that both describe the detrimental effects of global warming on the Arctic region, including the North-pole and the Greenland ice-sheet. Especially the melting of the latter will pose severe problems for countries at an elevation slightly above or even partly below current sea levels, like the Netherlands for example, possibly in a few decades from now.
2. This phenomenon has also been described as "micro-innovations" in Mokyr (1990), "refinements" in Jovanovic and Rob (1990), and "secondary innovations" in Aghion and Howitt (1998a, ch. 6).
3. Or, in the context of the family metaphor, its "founding father".
4. This combination of features was inspired by the focus of the original STAGEPOST model on identifying the combinations of technological characteristics that really matter for the growth of a single innovation into an entire GPT.
5. The underlying assumption is that the most important GPT extensions are made first. This in turn gives rise to decreasing returns to (applied) R&D.
6. The decreasing returns to variety are a relative novelty that hinges on the existence of asymmetries in the contribution of individual intermediates to their effective "ensemble". Such asymmetries are also present in van Zon (2001) and van Zon and Yetkiner (2003).
7. There are several reasons why an innovation may not grow into a full-fledged GPT. This may be due, for example, to high costs or a limited scope of expansion, or a low intrinsic productivity of the "core-idea". For more details, see section 2.2 below, and van Zon, Fortune and Kronenberg (2003) for further details.
8. Because new peripherals contribute less to GPT productivity than "older" peripherals by assumption, we actually have something looking like an inverted quality ladder.
9. It should be noted that Equations (1) and (2) provide a very simple structure that can easily be generalized. See van Zon, Fortune and Kronenberg (2003) for more details.
10. The inner organization of a GPT as a CES aggregate of a core and peripherals with Love of Variety features is in our case responsible for the "innovational complementarity" character of GPTs as advanced by Bresnahan and Trajtenberg (1995), who state that "the productivity of R&D in a downstream sector increases as a consequence of innovation in the GPT technology". In fact the notion of complementarity seems somewhat at odds with the interpretation of productivity increases due to improved division of production tasks between intermediates in a setting where individual intermediates are direct substitutes for each other. However, in our view the notion of complementarity in this context does not refer to the *necessity* of joint development but rather to its *desirability*.
11. This difference in the degree of uncertainty resembles Wälde (1999, 2004) in which both risky R&D and riskless capital accumulation activities are distinguished. In our model, however, the risky-ness of an activity does not directly influence decision making *ex ante* (although it could *ex post* through temporary lock-in, for example). Our differential uncertainty assumption serves the double purpose of simplicity and of highlighting the notion that finding a new GPT must be intrinsically more "difficult" than expanding upon an existing idea. It should be noted, however, that if we would incorporate risk aversion explicitly, this assumption would make basic research

less attractive compared to applied research. We will show, however, that a market failure will lead to an inefficient allocation between these two types of research anyway.

12. The expected present value as the incentive for basic R&D depends on the expected values of the characteristics of the core of the new GPT, but also on the expected development of marginal costs. For applied R&D, the characteristics are known once those of the core of the new GPT are known.
13. This effectively adds a Leontieff layer at the component level to the existing multilevel structure.
14. The names Ci and Nj refer to carbon-based GPT i and non-carbon-based GPT j.
15. Based on Equation (2), the "Eulerian" contribution of GPT z to the aggregate stock of effective capital K_e is calculated as $z \cdot (\partial K_e/\partial z)/K_e$.
16. In the real world, such a massive refocusing of R&D efforts would take more time, since R&D resources are not perfectly mobile as they are in the model, and information about new GPTs may take some time to diffuse. One might account for sluggish R&D labour movement in the model, but this would only complicate matters and divert attention away from the central issues.
17. It should be noted that the curve for applied R&D is really the horizontal summation over the VMP curves associated with applied R&D in all currently existing GPTs.
18. It should be noted that the introduction of two R&D sectors also leads to some superficial changes in Equations (12.A) and (12.B) in particular. The latter would now have to be respecified for both R&D sectors. In addition, we have to add an equation like (11) that describes the equalization of the value marginal product of R&D workers in basic research both in the carbon-sector and the non-carbon R&D sector. The full employment condition of R&D workers as given by Equation (12.A) also has to be reformulated so as to include employment in all different uses of R&D workers in both sectors. The logic of the allocation problem remains as straightforward as before, however, and so the adjusted equations are not listed here.
19. This is because the pervasiveness of a GPT depends for a large part on its performance relative to other existing GPTs. If, for some historical reason, the latter are not very productive, then even a mediocre GPT may become pervasive, in accordance with the saying "in the kingdom of the blind, the one-eyed man is king".
20. We use the first 100 "years" to get rid of the influence of the initial values of the capital–stock variables in the model.
21. However, we are now also working on an extension of the model that includes a simplified version of putty-clay capital. See van Zon (2005).

References

Aghion, P. and P. Howitt (1992), 'A Model of Growth through Creative Destruction', *Econometrica* **60**, 323–351.

Aghion, P. and P. Howitt (1998a), *Endogenous Growth Theory*. Cambridge MA: MIT Press.

Aghion, P. and P. Howitt (1998b), 'On the Macroeconomic Effects of Major Technological Change', *Annales d'economie et de statistique*, 49/50, 50–75.

Arrow, K. J. (1962), 'The Economic Implications of Learning by Doing', *Review of Economic Studies* **29**(3), 155–173.

Barro, R. J. and Sala-I-Martin, X. (1995), *Economic Growth*, New York: McGraw-Hill.

Bresnahan, T. and Trajtenberg, M. (1995), 'General Purpose Technologies', *Journal of Econometrics* **65**, 83–108.

Carlaw, K. I. and R. G. Lipsey (2001), 'Externalities versus Technological Complementarities: a Model of GPT-driven, Sustained Growth', paper presented at the Conference in Honour of the 20th Anniversary of Nelson and Winter's Book *An Evolutionary Theory of Economic Change*, http://www.sfu.ca/~rlipsey

Dosi, G. (1988), 'Sources, Procedures, and Microeconomic Effects of Innovation', *Journal of Economic Literature* (September), **26**, 1120–1171.
Ethier, W. J. (1982), 'National and International Returns to Scale in the Modern Theory of International Trade', *American Economic Review* (June) **72**(3), 389–405.
Francois, P. and H. Lloyd-Ellis (2003), 'Animal Spirits Through Creative Destruction', *American Economic Review* **93**(3), 530–550.
Freeman, C. and C. Perez (1988), 'Structural Crises of Adjustment, Business Cycles and Investment Behaviour', in Dosi et al. eds., *Technical Change and Economic Theory*. Pinter Publishers.
Grossman, G. M. and E. Helpman (1991), *Innovation and Growth in the Global Economy*. Cambridge MA: MIT Press.
Helpman, E. (1998), *General Purpose Technologies and Economic Growth*. MIT Press.
Helpman, E. and M. Trajtenberg (1994), 'A Time to Sow a Time to Reap: Growth Based on General Pupose Technologies', *Center for Economic Policy Research, Discussion Paper No. 1080* (December), 1–32.
Jovanovic, B. and R. Rob (1990), 'Long Waves and Short Waves: Growth through Intensive and Extensive Search', *Econometrica*, **58**(6), 1391–1409.
Lipsey, R. and C. Bekar (1995), 'A Structuralist View of Technical Change and Economic Growth', in T. Courchene, ed., *Technology, Information and Public Policy*. Kingston Ontario: John Deutsch Institute for Study of Economic Policy.
Matsuyama, K. (1999), 'Growing through Cycles', *Econometrica* **67**, 335–347.
Mensch, G. (1979), *Stalemate in Technology: Innovations Overcome Depression*. Cambridge MA: Ballinger Publishing Company.
Mokyr, J. (1990), *The Lever of Riches*. Oxford University Press.
Mukerji, V. (1963), 'Generalized SMAC Function with Constant Ratios of Elasticity of Substitution', *Review of Economic Studies* **30**, 233–236.
Olsson, O. (2001), 'Why Does Technology Advance in Cycles?', Available at: http://swopec.hhs.se/gunwpe/abs/gunwpe0038.htm.
Rifkin, J. (2002), *The Hydrogen Economy*. Cambridge: Polity.
Romer, P. (1990), 'Endogenous Technological Change', *Journal of Political Economy* **98**, 71–102.
Schumpeter, J. (1934), *The Theory of Economic Development*. Cambridge MA: Harvard University Press.
Schumpeter, J. (1939), *Business Cycles: A Theoretical, Historical, and Statistical Analysis of the Capitalist Process*, 2 Vols: New York: McGraw-Hill.
Segerstrom, P. S., T. Anant, and E. Donopoulos (1990), 'A Schumpeterian Model of the Product Life Cycle', *American Economic Review* **80**, 1077–1092.
Smulders, S. and de Nooij, M. (2002), 'The Impact of Energy Conservation on Technology and Economic Growth', *Resource and Energy Economics* **25**, 59–79.
Solow, R. (1997), *Learning from "Learning by Doing" Lesson for Economic Growth*. Stanford CA: Stanford University Press.
Solow, R. (2000), *Growth Theory: An Exposition*, Oxford: Oxford University Press.
Wälde, K. (1999), 'Optimal Saving under Poisson Uncertainty', *Journal of Economic Theory* **87**, 194–217.
Wälde, K. (2004), 'Endogenous Growth Cycles', downloaded from http://www.ires.ucl.ac.be/DP/IRES_DP/2004-12.pdf , forthcoming in the *International Economic Review*.
Yetkiner, I. H. (2003), *Shocks and Growth: Four Essays*, Ph.D. thesis, Groningen: Groningen University.
Zon, A. van (2001), 'A Simple Endogenous Growth Model with Asymmetric Employment Opportunities by Skill', *MERIT Research Memorandum* 2001–029.

Zon, A. van, E. Fortune, and T. Kronenberg (2003) 'How to Sow and Reap as You Go: a Simple Model of Cyclical Endogenous Growth', *MERIT Research Memorandum*, 2003–026.

Zon, A. van and I. H. Yetkiner (2003), 'An Endogenous Growth Model à la Romer with Embodied Energy-Saving Technological Change', *Resource and Energy Economics* **25**(1), 81–103.

Zon, A. van (2005) 'Vintage Modelling for Dummies Using the Putty-Practically-Clay Approach', *MERIT Research Memorandum*, 2005–005.

6. Efficient Dynamic Pollution Taxation in an Uncertain Environment

SUSANNE SORETZ
Department of Economics, University of Hannover, Königsworther Platz 1, Hannover, D-30 167, Germany; Author for correspondence (e-mail: soretz@vwl.uni-hannover.de)

Abstract. This paper analyzes efficient pollution taxation within a stochastic model of endogenous growth. Pollution is a by-product of production and causes disutility. Furthermore, the productivity which results from environmental quality is uncertain. This reflects e.g. uncertain capital depreciation induced by natural disasters like hurricanes or floods. This uncertainty is shown to raise an ambiguous impact on the optimal pollution level as well as on optimal environmental taxation. Market equilibrium turns out to be suboptimal, since the households mis-perceive their individual impact on pollution. Conditions for welfare maximizing pollution taxation are stated and it is shown that a direct pollution tax is not appropriate to yield Pareto-optimal growth. Instead, a linear capital income tax together with a linear abatement subsidy build an efficient tax scheme, if secondarily the governmental budget is balanced. Moreover, an increase in the riskiness of environmental productivity may even lead to an increase in the optimal pollution level and to a decrease in optimal environmental taxation, depending predominantly on the preference parameters.

Key words: pollution, taxation, uncertainty, endogenous growth

1. Introduction

The main question of this paper is how a society should react to environmental risk. Given that we do not know the detailed consequences of global warming on our future well being: should we aspire less carbon dioxide emissions due to the risk as a precaution? Should we increase the pollution tax as a response to risk? I develop a model of a growing economy subject to pollution externalities and uncertainty. The analysis of the model shows that there are no unambiguous answers to the questions raised above. When uncertainty about the link between pollution and production is bigger, the optimal level of pollution may become larger and the optimal pollution tax smaller. The former is due to the rise in optimal capital accumulation which can result from the increase in risk. The latter is caused by the increase in equilibrium abatement activity going back to the extended risk associated with pollution.

This paper discusses the dynamic situation with pollution as a by-product of production and with environmental risk: Environmental quality is assumed to have an impact on productivity, but this impact is uncertain. The number of employees away sick increases due to environmental degradation, but the number away sick

L. Bretschger and S. Smulders (eds.), Sustainable Resource Use and Economic Dynamics, 101–125.

in a specific firm is not known with certainty in advance. Environmental degradation also increases the destructive power of hurricanes or floods. Nevertheless, it is uncertain which firm will be affected by unforeseen depreciation. The main focus of the paper is to develop welfare maximizing pollution taxation in a dynamic and uncertain world. Many contributions analyze environmental policies within a static, riskless setting. I demonstrate that the optimality of environmental tax schemes does not necessarily carry over to the stochastic dynamic setting. In particular, a tax which is levied directly on pollution, is shown to be inconsistent with steady state growth. Additionally, as long as the considered risk is idiosyncratic, a welfare maximizing pollution tax provides an insurance against the income volatility caused by environmental risk and thereby completely eliminates income uncertainty. The pollution level reacts ambiguously to the abolition of risk, depending on the parameter setting.

Various contributions examine the effect of environmental degradation on endogenous growth, as e.g. Gradus and Smulders (1993), Ligthart and van der Ploeg (1994), Bovenberg and Smulders (1997), Jones and Manuelli (1995), Byrne (1997) or Stokey (1998). The authors derive conditions for the existence of sustainable growth paths and analyze environmental policy to internalize market failures. In general, the effect of environmental aspects on the growth process differs with respect to the underlying production structure. If environmental degradation is an inevitable by-product of the consumption good, as e g. in the approach of Stokey (1998), sustainable growth is unlikely due to the trade-off between consumption and environment. In contrast, if the engine of growth is independent from environment, as e.g. the accumulation of human capital in the Lucas-type model of Gradus and Smulders (1993) or Byrne (1997), the optimal (sustainable) growth path may even be unaffected by environmental concerns. Hartman and Kwon (2005) show in a related setting, that an environmental Kuznets curve might occur.

In the presence of pollution, uncertainty is important, since risk averse individuals will adjust their decisions to the underlying uncertainty. Although risk is a determining factor of the evolution of environmental quality as well as of the growth process, there are only few papers which address the impact of risk on environmental development, as e.g. Baranzini and Bourguignon (1995), Beltratti (1998), Chichilnisky and Heal (1998) as well as Ayong Le Kama and Schubert (2004) or Keller et al. (2004). Uncertainty gains importance for the dynamic macroeconomic equilibrium mainly through two different ways: On the one hand side, risk averse individuals react on the underlying risk within their intertemporal decision concerning consumption, abatement, and capital accumulation. The reaction is ambiguous and depends crucially on intertemporal substitution as well as on risk aversion (see e.g. Soretz 2003, 2004). On the other hand side, the impact of environmental policy changes due to uncertainty. Any governmental activity influences not only expected values of net economic variables, but also their volatility. This leads to counter acting effects on the equilibrium growth process, which were analyzed first in the seminal work of Eaton (1981) and more recently taken up e.g.

by Turnovsky (1993, 1995b, 2000), Smith (1996a), Clemens and Soretz (1997, 2004), or Corsetti (1997).

The usual assumption of environmental quality as a pure public good is relaxed in this model. Instead, the agents take part of their influence on pollution into account within individual optimization: With respect to some aspects, environmental quality exhibits rivalry. For instance, vegetables which are cultivated without pesticides, are healthful for the particular consumer *and* for the whole society. The extent to which pollution is perceived to be unattached to individual decisions is parameterized according to the formulation of congestion effects in the public goods literature (see e.g. Edwards 1990, Glomm and Ravikumar 1994; Fisher and Turnovsky 1998; Turnovsky 1999). Nevertheless, due to this partial perception, market equilibrium is suboptimal, and gives the reason for pollution taxation.

The main focus of this paper is the following: In a stochastic environment which emphasizes the uncertainty of the productivity due to environmental quality, conditions for efficient pollution taxation are developed. It is shown that due to uncertainty, a tax which is levied directly on pollution, is not suitable to obtain socially optimal growth. However, a linear tax on capital together with a linear subsidy on abatement and a balanced governmental budget is a simple example for welfare maximizing pollution taxation. If the uncertainty associated with pollution is idiosyncratic, government is able to provide an insurance against the involved income volatility. Hence, a complete insurance results to be required for a first best pollution tax in a society of risk averse individuals. Nevertheless, the first best growth rate as well as the first best sustainable pollution level react ambiguously to the elimination of risk.

The assumptions of the model are presented in Section 2. Section 3 develops the Pareto-optimal growth path to serve as reference setting. Part 4 derives the decentralized equilibrium. Section 5 establishes conditions for efficient environmental policy and analyzes different types of pollution taxation. Section 6 gives a short conclusion.

2. The Model

According to the formulation of Smulders and Gradus (1996), environmental quality affects the economy through various channels: first, pollution is an inevitable by-product of production. Second, environmental quality affects the productivity within the consumption good sector. Third, pollution causes disutility. These three effects will be defined in the following.

In the underlying economy pollution is caused by capital accumulation and reduced by means of abatement effort. Hence, within the growth process, pollution increases as a by-product whenever the households invest in physical capital. Pollution decreases when households decide to raise abatement expenditures. Smulders and Gradus (1996) show that sustainable growth with non-increasing long-run pollution in this setting only is feasible if the elasticity of pollution with respect to abatement is greater or equal than the elasticity of pollution with respect to capital. Roughly speaking, pollution is non-increasing during the growth process, if abatement is at least as effective than capital with respect to pollution. Here, the limiting case with equal elasticities will be considered. Without loss of generality,[1] individually caused pollution, $P_i(t)$, is simply defined by the ratio between the individual capital stock, $k_i(t)$, and individual abatement effort, $e_i(t)$,

$$P_i(t) = P(k_i(t), e_i(t)) = \frac{k_i(t)}{e_i(t)} \tag{1}$$

such that the elasticities of pollution with respect to capital and abatement are equal (and unity).[2]

Furthermore, pollution is considered to be a flow variable, hence the model can predominantly be applied to pollutants which dissolve rather quickly. Nevertheless, this assumption seems maintainable since the pollution level is linked to the stock of physical capital. Therefore, ongoing capital accumulation ceteris paribus induces a perpetual increase in pollution.

The second effect of pollution refers to the productivity within the consumption good sector. The consumption good is produced by the only input factor capital and by means of a linear technology. In order to keep the framework as simple as possible, labor is neglected. Hence, $k_i(t)$ should be interpreted as broad measure of capital, including human capital. The production function of household i follows Smulders and Gradus (1996) or Stokey (1998) and can be written as

$$f_i(k_i(t)) = Ap_i(t)k_i(t) \tag{2}$$

where productivity $Ap_i(t)$ depends on environmental quality. An increase in environmental quality raises productivity e.g. by reducing depreciation of physical capital or by enhancing the health of workers (see Smulders and Gradus 1996, p. 508). Moreover, the productivity impact of the environment, $p_i(t)$, is not known with certainty in advance. A reason for this assumption is that uncertainty is a main feature of environmental degradation: Low environmental quality e.g. increases the power and the quantity of hurricanes which cause capital depreciation. Nevertheless, the occurrence of hurricanes is rather stochastic than deterministic. It is not known with certainty, when the next hurricane will occur and which factory it will destroy.

Therefore, only the expected value of environmental quality is known, but additionally there is environmental uncertainty: The productivity effect of environmental quality is stochastic and determined by[3]

$$p_i(t) = P(t)^{-\alpha}dt + P(t)^{\alpha'}\sigma dz_i(t)\alpha, \quad \alpha' > 0 \tag{3}$$

where $dz_i \sim N(0, dt)$ denotes the individual specific increment to a Wiener process.[4] Since environmental quality is a public good, aggregate pollution, $P(t)$, is relevant for environmental quality and therefore determines the expected productivity effect of pollution. With the assumption of a continuum of individuals with homogenous preferences as well as homogenous technology, it will be derived subsequently, that all individuals emit the same amount of pollution.

Additionally, population size is normalized to unity,[5] so the aggregate will be described by the average. Therefore, aggregation of the individual pollution levels ends up in the identity of aggregate (average) and individual pollution. With the definition given in (3), α determines the absolute value of the elasticity of expected production with respect to pollution. Hence $\alpha > 1 (0 < \alpha < 1)$ means that an increase in pollution by 1 percent ceteris paribus results in a decrease in expected production by more (less) than one percent. This is equivalent to an increasing (decreasing) marginal productivity of environmental quality. The case $\alpha = 1$ indicates a constant marginal product of environmental quality. From the empirical point of view, none of these cases can be excluded. The value of α depends on the considered industry.

A natural disaster, e.g. the occurrence of a hurricane or a flood, is represented by a low (or negative) realization of the stochastic disturbance, dz_i, which results in a low productivity. With this definition, the standard deviation of environmental productivity, $P_i^{\alpha'} \sigma dt$, increases with the aggregate pollution level. This assumption reflects the fact that lower environmental quality comes along with an increase in the destructive power of hurricanes or an increase in the water level of floods. Therewith the absolute value of environmental degradation which results from natural disasters increases with the mean level of pollution used in production.[6] This assumption resembles the settings of Fernandez (2005) or Lafforgue (2005).

The features of the stochastic process, dz_i, (the expected value as well as the riskiness) are equal for all firms. Nevertheless, the realization of the environmental risk is individual for each firm. The reason is that this paper focuses on idiosyncratic environmental risk, as e.g. the number of staff away sick due to environmental reasons, rather than aggregate risk, which affects the whole society, as e.g. accidents in a nuclear power station. Of course, in reality most types of environmental risk are neither idiosyncratic nor aggregate. The probabilities that different factories will be destroyed by a hurricane rather are positively correlated. Nevertheless, hurricanes or floods usually emerge locally, hence in the following they are considered as approximately idiosyncratic. Moreover, the inclusion of correlated environmental risk to the point of aggregate risk does not imply major changes to the results.

The utility of individual i depends on his consumption path, $c_i(t)$, as well as on the aggregate pollution path, $P(t)$. Furthermore, the individuals are infinitely long lived[7] and their intertemporal utility is defined according to the recursion

$$\begin{aligned} G((1-\rho)u_i(t)) &= \frac{1-\rho}{1-1/\varepsilon}(c_i(t)P(t)^{-\gamma})^{1-1/\varepsilon}h \\ &\quad + \exp(-\beta h)G((1-\rho)\mathrm{E}_t[u_i(t+h)]) \end{aligned} \tag{4}$$

$$\text{with } G_i = G(x_i) = \frac{1-\rho}{1-1/\varepsilon}x_i^{\frac{1-1/\varepsilon}{1-\rho}}, \quad \varepsilon \neq 1, \rho \neq 1. \tag{5}$$

The constant rate of time preference is denoted with $\beta > 0$, and $\gamma > 0$ indicates disutility out of pollution. With an increase in γ, environmental amenities gain importance.

Recursive preferences were applied to stochastic growth by Obstfeld (1994) and later for instance by Smith (1996b), in order to distinguish the effects of risk taking from those of intertemporal substitution. The recursive specification of intertemporal utility draws back on Epstein and Zin (1989) or Weil (1990) and was extended to continuous time by Svensson (1989) and Duffie and Epstein (1992). It allows for a constant intertemporal elasticity of substitution, $\varepsilon > 0$, as well as a constant degree of relative risk aversion, $\rho > 0$. Nevertheless, it is possible to set these two parameters separately. In the special case where $1/\varepsilon = \rho$, the recursive preferences (4) result in the usual expected utility form.

3. Pareto-optimal Growth

In this section, the socially optimal growth path will be determined. Therefore, a social planner chooses individual consumption and abatement expenditures together with capital accumulation in order to maximize lifetime utility, $u_i(t)$, as defined by the recursion (4) and (5), with respect to the capital accumulation process

$$dk_i = [Ak_i P^{-\alpha} - c_i - e_i]dt + Ak_i P^{\alpha'} \sigma dz_i. \tag{6}$$

The stochastic Bellman equation is derived by means of Itô's Lemma since the stochastic accumulation process of capital cannot be differentiated with respect to time. Additionally, let the value function, $J_i(k_i)$, denote maximum lifetime utility of individual i. The optimization problem of the social planner then results in the stochastic Bellman equation

$$\begin{aligned} \mathcal{B} = {} & \frac{1-\rho}{1-1/\varepsilon}(c_i P^{-\gamma})^{1-1/\varepsilon} - \beta G((1-\rho)J_i(k_i)) \\ & + (1-\rho)G'((1-\rho)J_i(k_i)) \left(J_i'(k_i)\frac{\mathrm{E}[dk_i]}{dt} + \frac{1}{2}J_i''(k_i)\sigma_{k_i}^2 \right) \end{aligned} \tag{7}$$

with given initial values of physical capital, k_{i0}, and of the stochastic disturbance, $z_{i0} = 0 \; \forall i$. The variance of individual capital, $\sigma_{k_i}^2 \equiv (E[dk_i^2] - E[dk_i]^2)/dt = A^2 k_i^2 P^{2\alpha'} \sigma^2$, is determined by the volatility of the productivity effect of environmental quality.

Maximization of the Bellman equation with respect to consumption leads to the first necessary condition

$$c_i^{-1/\varepsilon} P^{-\gamma(1-1/\varepsilon)} - G_i' J_i' \stackrel{!}{=} 0 \tag{8}$$

which balances marginal utility out of consumption across time. The assumptions of constant intertemporal elasticity of substitution, ε, and constant degree of relative risk aversion, ρ, together with an intratemporal elasticity between consumption and pollution which is unity lead to the following conjecture

$$\mu_i \equiv \frac{c_i}{k_i} = \mu \quad \forall i, t, \qquad \eta_i \equiv \frac{e_i}{k_i} = \eta \quad \forall i, t \tag{9}$$

of constant consumption and abatement ratios. Particularly, this means the independence of consumption and abatement ratios from the individual capital stock and implies homogeneity of all households with respect to these decisions, independent from the realization of their environmental productivity time path. Hence, consumption and abatement activities are presumed to grow at the same rate as physical capital in the socially optimal steady-state and this growth rate is supposed to be equal for all individuals. Then neither the distribution of initial endowment nor the time path of environmental productivity have an impact on expected growth. Growth as well as the optimal ratios (9) are equal for all individuals. Moreover, since population is normalized to unity, aggregate economic variables are identical with their mean values.

Together with the definition (5) of G_i, substitution of $P = \eta^{-1}$, and the necessary condition (8), this conjecture results in a CRRA guess for the value function

$$J(k_i) = \left(\mu\eta^{\gamma(1-\varepsilon)}\right)^{\frac{1-\rho}{1-\varepsilon}} \frac{k_i^{1-\rho}}{1-\rho}. \tag{10}$$

Maximization of the Bellman equation with respect to abatement expenditures, e_i, gives the second necessary condition

$$\gamma(c_i P^{-\gamma})^{1-1/\varepsilon} + G_i' \left(J_i'(\alpha A k_i P^{-\alpha} - e_i) + \frac{1}{2} J_i'' e_i \frac{\partial \sigma_{k_i}^2}{\partial e_i} \right) \stackrel{!}{=} 0 \tag{11}$$

which balances marginal utility out of consumption and abatement, and yields intratemporal effeciency. Substitution of the value function (10), the definition (5) of G_i and the constant ratios μ and η leads to the following relationship

$$\frac{\mu\gamma}{\eta} = 1 - \alpha A \eta^{\alpha-1} \left(1 + \rho \frac{\alpha'}{\alpha} A \eta^{-2\alpha'-\alpha} \sigma^2\right) \tag{12}$$

which equates the marginal rate of (intratemporal) substitution, dc/de, with the price relation, and determines optimal consumption and abatement expenditures. The right hand side of Equation (12) gives the certainty equivalent of the marginal cost of an increase in abatement (expressed in units of the consumption good): first, there is a direct cost of one unit, and second, there is an indirect negative cost since environmental quality raises the productivity of the consumption good sector.[8] The certainty equivalent of the marginal cost of an increase in abatement, expressed in units of the consumption good, is equated with the marginal rate of substitution, dc/de, which is given on the left hand side of Equation (12). Hence, the negative impact of pollution on output ($-\alpha$) ceteris paribus increases the optimal abatement ratio, compared with a situation where pollution had no effect on production ($\alpha = 0$). This replicates the additional benefit of abatement activity due to enhanced productivity.

Furthermore, uncertainty (positive σ^2) ceteris paribus decreases the optimal abatement ratio. With more abatement activity, pollution decreases and therefore the volatility of environmental productivity diminishes: if a cleaner production process is chosen, there is less risk of natural disasters which reduce productivity.

The more risk averse the households are (higher ρ), the stronger is the socially optimal reaction on environmental risk. Nevertheless, the entire impact of uncertainty on the socially optimal pollution level is more complex and turns out to be ambiguous. The relationship (12) contains the optimal consumption ratio, μ, as well as the optimal abatement ratio, η. Both will be determined subsequently together with Equation (14) and react ambiguously on environmental risk, as explained in more detail with respect to the optimal growth rate (16).

Maximization of the stochastic Bellman Equation (7) with respect to capital leads to the third necessary condition

$$-\gamma(cP^{-\gamma})^{1-1/\varepsilon}k_i^{-1} + (1-\rho)G_i''J_i'\left(J_i'\frac{\mathrm{E}[dk_i]}{dt} + \frac{1}{2}J_i''\sigma_{k_i}^2\right)$$
$$+G_i'\left(J_i'(AP^{-\alpha}(1-\alpha)-\beta) + J_i''\left(\frac{\mathrm{E}[dk_i]}{dt} + \frac{1}{2}\frac{\partial\sigma_{k_i}^2}{\partial k_i}\right) + \frac{1}{2}J_i'''\sigma_{k_i}^2\right) \stackrel{!}{=} 0 \qquad (13)$$

which determines optimal capital accumulation and weighs momentary utility out of consumption or abatement against future utility out of capital investment. Again, substitution of (5), (10) and (9) together with the optimal abatement decision (12) results in the optimal consumption ratio

$$\mu = \varepsilon\beta + (1-\varepsilon)\left(A\eta^{\alpha} - \eta - \frac{\rho}{2}A^2\eta^{-2\alpha'}\sigma^2\right). \qquad (14)$$

An increase in capital accumulation yields the expected social return $A\eta^{\alpha} - \eta$, and simultaneously increases the riskiness of future income. Whether a rise in capital returns leads to a positive or a negative effect on consumption, depends on the elasticity of intertemporal substitution, ε, to be less or greater than unity. Together with relation (12), the socially optimal consumption and abatement ratios are determined. Condition (14) focuses on the dynamic trade-off between momentaneous consumption (or abatement) and capital accumulation, whereas Equation (12) describes the intratemporal trade-off between consumption and abatement.

Unfortunately, it is not straightforward to determine a closed-form solution. Inserting Equation (14) into (12) leads to the polynomial

$$1 - \alpha A\eta^{\alpha-1}\left(1 + \rho\frac{\alpha'}{\alpha}A\eta^{-2\alpha'-\alpha}\sigma^2\right)$$
$$= \frac{\gamma}{\eta}\left(\varepsilon\beta + (1-\varepsilon)\left(A\eta^{\alpha} - \eta - \frac{\rho}{2}A^2\eta^{-2\alpha'}\sigma^2\right)\right) \qquad (15)$$

determining optimal abatement activity. Without further specification, it is impossible to calculate the optimal level of abatement expenditure.[9] Following, the solution will be based on the interpretation of condition (15). This condition equates marginal cost of abatement (left hand side) with marginal benefit of abatement (right hand side). The curves of marginal cost and marginal benefit can be shown to intersect uniquely[10] as long as the elasticity of intertemporal substitution is sufficiently low, $\varepsilon < 1$, which is the empirically relevant range (see e.g. Hall 1988; Epstein and Zin 1991). Figure 1 demonstrates the unique optimal solution for abatement effort.

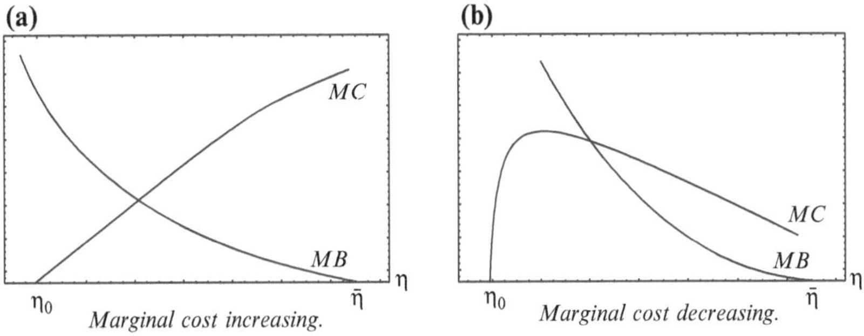

Figure 1. Optimal abatement effort.

Now, from goods market clearing, $dk = f - c - e$, it is straightforward to determine the expected capital growth rate

$$\varphi \equiv \frac{E[dk]}{kdt} = \varepsilon(A\eta^{\alpha} - \eta - \beta) + (1 - \varepsilon)\frac{\rho}{2}A^2\eta^{-2\alpha'}\sigma^2 \tag{16}$$

which describes socially optimal growth, given the optimal abatement ratio from Equation (15). Optimal expected growth, as well as optimal consumption and abatement indeed are independent from the capital stock and hence confirm the conjecture of equal and constant growth rates of all economic variables. Through optimal abatement effort, η, the social marginal return on capital is reduced and hence optimal expected growth falls short of the growth rate obtained with an AK-technology without pollution.

Furthermore, one can see that environmental risk affects socially optimal growth in an ambiguous way: First, due to the riskiness of environmental productivity, the certainty equivalent of capital return (unambiguously) decreases. Now, as well known, the optimal reaction on this decrease in the capital return depends on the elasticity of intertemporal substitution. If it is sufficiently low ($\varepsilon < 1$), the income effect dominates and leads to an increase in savings in order to compensate for the reduction in capital return. If instead the elasticity of intertemporal substitution is sufficiently high ($\varepsilon > 1$), optimal capital accumulation is reduced in favor of momentaneous consumption and abatement, due to the dominance of the substitution effect.[11]

Note that sustainable growth is feasible in the underlying economy: If social capital productivity is sufficiently high to allow for a positive growth rate in any time increment, expected growth will continue for all time increments. Additionally, optimal pollution remains constant since optimal abatement grows with the same (stochastic) rate as capital. Nevertheless, the impact of the environmental productivity risk on the optimal pollution level and the optimal growth path is not clear cut: The optimal response on an increase in the riskiness may consist of an increase in the pollution level together with an increase in growth, due to the dominance of the income effects.

Last, the transversality condition

$$\lim_{t\to\infty} \mathrm{E}\left[\exp(-\beta t)G((1-\rho)J(k_i))\right] = 0 \tag{17}$$

must be satisfied in order to assure the boundedness of the optimal solution.[12]

4. Dynamic Market Equilibrium

One major problem of environmental degradation is that usually there are externalities which lead to suboptimal outcomes of market solutions. In the following two sections, the goal is to describe optimal pollution taxation in the stochastic dynamic setting. Therefore, this section will demonstrate the specific failures of market equilibrium: There is a gap between private and social marginal costs of pollution. Households – being firms at the same time – underestimate their individual impact on aggregate pollution. It will be shown that in this setting only the intratemporal decision between consumption and abatement is disturbed, whereas the intertemporal decision about capital accumulation corresponds to the Paretooptimum.

The environment is often assumed to be a public good. Hence, individuals free ride and do not take into account their individual impact on pollution at all. For two reasons, this assumption will be relaxed in the further analysis. First, there are various environmental goods, which display rivalry to some degree: for example, food or clothes which are ecologically compatible, on the one hand side protect the individual health (only of the individual who bought them), and on the other hand side protect the environment (of the whole society). Second, persons show to some degree morality with respect to the environment: Sometimes, people go by bike to the bakery (and not by car), because they know that this behavior reduces air pollution for the whole society, and for themselves, too. Accordingly, Eriksson (2002, p. 281) states that households are "[...] willing to pay an extra premium for a product if it were green".

Hence, the perception of the individual influence on pollution is parameterized: I do not assume a pure pollution externality, where individuals do not consider pollution at all. But the individuals neither feel completely responsible for their individual impact on pollution. In fact, they perceive pollution to depend in part, δ, on the decisions of "the others", hence on aggregate (average) capital accumulation, k, and aggregate (average) abatement effort, e, which are exogenous to individual decisions. Only the (maybe very small) part $1-\delta$ of pollution is perceived to depend on "the own" behavior, hence on individual capital accumulation, k_i, and individual abatement expenditures, e_i. Perceived pollution, P_p, is then given by

$$P_\mathrm{p} = \left(\frac{\mathrm{k(t)}}{e(t)}\right)^{\delta} \left(\frac{k_i(t)}{e_i(t)}\right)^{1-\delta}, \quad \delta \in [0, 1) \tag{18}$$

and replaces pollution within utility and production functions for individual optimization.

This setting of perception relies on the formulation of congestion effects in the public goods literature (see e.g. Edwards 1990; Glomm and Ravikumar 1994; Turnovsky 1999). In these lines, $1 - \delta$ can be interpreted as degree of locality of pollution. The part δ of pollution is global and affects the whole society in the same way. The part $1-\delta$ of pollution is local and affects only the individual which caused the pollution. With this respect, $1-\delta$ is the joint degree of rivalry of capital and abatement in the "production" of pollution (see Turnovsky 1995, p. 405).

As long as the perception parameter is above zero, the agents underestimate their individual influence on pollution. The result is a negative externality of capital accumulation and a positive externality of abatement effort.

Subsequently will be shown that in market equilibrium the abatement ratios are equal for all individuals. Hence, the aggregate relation k/e and the individual relation k_i/e_i in perceived pollution (18) will be equal in equilibrium. Nevertheless, there is a continuum of individuals and the interaction between them is characterized by perfect competition. Therefore, they consider aggregate capital as well as aggregate abatement as exogenous within individual optimization.

Hence, the consumption and abatement ratios in market equilibrium have to fulfill[13]

$$1-(1-\delta)\alpha A\eta_{\mathrm{M}}^{\alpha-1}\left(1+\rho\frac{\alpha'}{\alpha}A\eta_{\mathrm{M}}^{-2\alpha'-\alpha}\sigma^{2}\right)$$
$$=\frac{\gamma(1-\delta)}{\eta_{\mathrm{M}}}\left(\varepsilon\beta+(1-\varepsilon)\left(A\eta_{\mathrm{M}}^{\alpha}-\eta_{\mathrm{M}}-\frac{\rho}{2}A^{2}\eta_{\mathrm{M}}^{-2\alpha'}\sigma^{2}\right)\right). \tag{19}$$

In comparison with the corresponding condition (15) for Pareto-optimal abatement effort, marginal cost (given on the left hand sides of 15 and 19) increases and marginal benefit decreases due to mis-perception, $1 - \delta$, as can be seen in Figure 2. Therefore, individually optimal abatement is reduced. This displays the externality due to partial perception of the individual impact on pollution. The households take only the part $1 - \delta$ of their individual influence on pollution into account. Hence, they only perceive part of their disutility out of pollution $(\gamma\mu_{\mathrm{M}})$ as well as part of the marginal environmental productivity $\left(\alpha A\eta_{\mathrm{M}}^{-\alpha-1}\right)$. Note that individuals also take only part of their impact on environmental productivity risk into account. Hence, the positive effect of pollution uncertainty on equilibrium abatement effort is less than the effect on Pareto-optimal abatement.

The marginal benefit of abatement given on the right hand side of Equation (19) is derived from the intertemporal decision between momentaneous consumption or abatement and future consumption through capital accumulation. This condition exactly replicates the corresponding condition for Pareto-optimal consumption. Since momentaneous consumption and future consumption are misperceived in the same way, there is no distortionary effect of partial perception on the intertemporal decision about capital accumulation.[14] Calculation of expected equilibrium growth out of market clearing $dk = f - c - e$ again displays this result

$$\varphi_{\mathrm{M}}=\varepsilon(A\eta_{\mathrm{M}}^{\alpha}-\eta_{\mathrm{M}}-\beta)+(1-\varepsilon)\frac{\rho}{2}A^{2}\eta_{\mathrm{M}}^{-2\alpha'}\sigma^{2}. \tag{20}$$

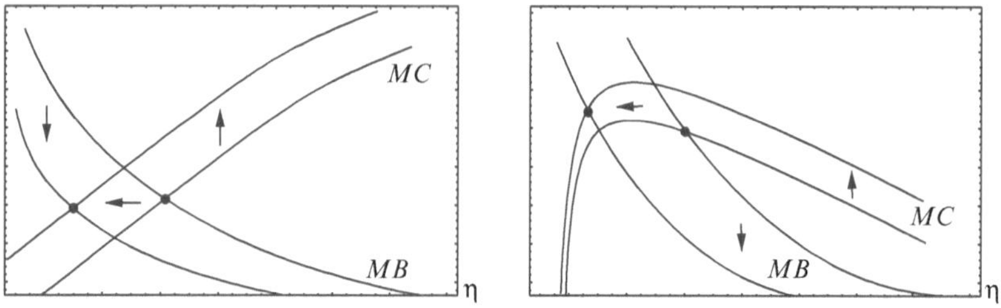

Figure 2. Impact of mis-perception on market equilibrium.

Note, that the conformity of intertemporal choice (20) with Pareto-optimal capital accumulation (16) does not imply that the expected growth rate in market equilibrium is Pareto-optimal. Due to partial perception, the equilibrium abatement ratio is suboptimally low, as already shown with Figure 2. Therefore, expected growth in market equilibrium is suboptimal. It is now straightforward to show that equilibrium expected growth is too high. The growth rate increases with a decline in the abatement ratio

$$\frac{\partial \varphi_{\mathrm{M}}}{\partial \eta_{\mathrm{M}}} = -\left(1 - \varepsilon\alpha A\eta_{\mathrm{M}}^{\alpha-1}\left(1 + \rho\frac{\alpha'}{\alpha}A\eta_{\mathrm{M}}^{-2\alpha'-\alpha}\sigma^2\right)\right) - \alpha'\rho A^2\eta_{\mathrm{M}}^{-2\alpha'-1}, \quad \sigma^2 < 0. \tag{21}$$

With a suboptimally low equilibrium abatement ratio (due to the positive externality of abatement effort), the equilibrium growth rate results to be suboptimally high (due to the negative externality of capital accumulation). This result gives rise to the introduction of pollution taxation: By means of a pollution tax, the true benefits and costs of pollution can be carried over to the households and market equilibrium can be improved.

5. Efficient Pollution Taxation

In order to internalize the externalities, a pollution tax will be incorporated. It will be demonstrated that with efficient pollution taxation, it is possible to realize Pareto-optimal equilibrium growth. Nevertheless, efficient pollution taxation implies distinct tax rates on capital and abatement on the one hand side, and different tax rates on deterministic and stochastic income components, on the other hand side.

The starting point is a generally formulated pollution tax

$$T_i(t) = T^{\mathrm{d}}(k_i(t), e_i(t))dt + T^{\mathrm{s}}(k_i(t), e_i(t))\sigma dz_i \tag{22}$$

which allows for a differentiated treatment of capital and abatement as well as of deterministic and stochastic income parts, and will be determined efficiently in equilibrium. With this pollution tax, the evolution of the capital stock becomes

$$dk_i = \left[Ak_iP_{\mathrm{p}}^{-\alpha} - c_i - e_i - T^{\mathrm{d}}(k_i, e_i)\right]dt + (Ak_iP_{\mathrm{p}}^{\alpha'} - T^{\mathrm{s}}(k_i, e_i))\sigma dz_i \tag{23}$$

and the variance of capital is given by

$$\sigma_{k_i}^2 = (A^2 k_i^2 P_{\mathrm{p}}^{2\alpha'} - 2Ak_i P_{\mathrm{p}}^{\alpha'} T^{\mathrm{s}} + (T^{\mathrm{s}})^2)\sigma^2. \tag{24}$$

As subsequently will be shown, any optimal governmental policy requires a balanced governmental budget where capital tax revenues equate abatement subsidy payments. Therefore, it is most convenient to focus only on pollution taxation as given in Equation (22) and to neglect further governmental expenditures.

Macroeconomic equilibrium depends twofold on pollution taxation: First, the intratemporal decision between consumption and abatement is influenced by the pollution tax. Second, within the intertemporal savings decision of risk averse individuals, they react on the environmental policy. The influence of pollution taxation on the equilibrium abatement ratios results in

$$\begin{aligned}
&\frac{1}{1-\delta}\left(1 + T_{e_i}^{\mathrm{d}} - \alpha(1-\delta)A\eta_{\mathrm{T}}^{\alpha-1}\right.\\
&\qquad\left. -\rho\sigma^2\left(\alpha'(1-\delta)A\eta_{\mathrm{T}}^{-\alpha'-1} - T_{k_i}^{\mathrm{s}}\right)\times\left(A\eta_{\mathrm{T}}^{-\alpha'} - \frac{T^{\mathrm{s}}}{k_i}\right)\right)\\
&= \frac{\gamma}{\eta_{\mathrm{T}}}\left(\varepsilon\beta + (1-\varepsilon)(A\eta_{\mathrm{T}}^{\alpha} - \eta_{\mathrm{T}}) - \left(\frac{T^{\mathrm{d}}}{k_i} - \varepsilon\left(T_{k_i}^{\mathrm{d}} + \eta_{\mathrm{T}}T_{e_i}^{\mathrm{d}}\right)\right)\right.\\
&\qquad + \frac{\rho}{2}\sigma^2\left((\varepsilon-1)A^2\eta_{\mathrm{T}}^{-2\alpha'} + 2A\eta_{\mathrm{T}}^{-\alpha'}\left(\frac{T^{\mathrm{s}}}{k_i} - \varepsilon(T_{k_i}^{\mathrm{s}} + \eta_{\mathrm{T}}T_{e_i}^{\mathrm{s}})\right)\right.\\
&\qquad\left.\left. - \left((1+\varepsilon)\frac{T^{\mathrm{s}}}{k_i} - 2\varepsilon(T_{k_i}^{\mathrm{s}} + \eta_{\mathrm{T}}T_{e_i}^{\mathrm{s}})\right)\frac{T^{\mathrm{s}}}{k_i}\right)\right).
\end{aligned} \tag{25}$$

Different from mis-perception, δ, the pollution taxation acts on the marginal cost of abatement (left hand side) and on the marginal benefit (right hand side) manifold. As long as the pollution tax is not established concretely, the impact on macroeconomic equilibrium is not clear cut. Nevertheless, it is now possible to determine equilibrium growth

$$\begin{aligned}
\varphi_{\mathrm{T}} &= \varepsilon\left(A\eta_{\mathrm{T}}^{\alpha} - T_{k_i}^{\mathrm{d}} - \eta_{\mathrm{T}}\left(1 + T_{e_i}^{\mathrm{d}}\right) - \beta\right)\\
&\quad - \frac{\rho}{2}\sigma^2\left((\varepsilon-1)A^2\eta_{\mathrm{T}}^{-2\alpha'} + 2A\eta_{\mathrm{T}}^{-\alpha'}\left(\frac{T^{\mathrm{s}}}{k_i} - \varepsilon\left(T_{k_i}^{\mathrm{s}} + \eta_{\mathrm{T}}T_{e_i}^{\mathrm{s}}\right)\right)\right.\\
&\quad\left. - \left((1+\varepsilon)\frac{T^{\mathrm{s}}}{k_i} - 2\varepsilon\left(T_{k_i}^{\mathrm{s}} + \eta_{\mathrm{T}}T_{e_i}^{\mathrm{s}}\right)\right)\frac{T^{\mathrm{s}}}{k_i}\right).
\end{aligned} \tag{26}$$

Taxation has both a direct and an indirect impact on equilibrium capital accumulation. The direct effect can be seen in Equation (26). The indirect impact is due to the adjustment of abatement effort. Within the direct influence, the pollution tax can again be shown to affect equilibrium growth through various channels: First, the expected returns on capital and abatement change. This can be seen in the first parenthesis of the growth rate (26). Expected capital return decreases due to the introduction of the pollution tax $\left(T_{k_i}^{\mathrm{d}}\right)$ and leads to the usual growth diminishing effect of distortionary capital income taxation. A common pollution tax will decrease in abatement effort $\left(\eta_T T_{e_i}^{\mathrm{d}} < 0\right)$ and in this case foster equilibrium growth. Expected return on abatement increases and therefore facilitates capital accumulation.

Second, the stochastic pollution tax affects the uncertainty associated with environmental quality. The reaction of risk averse individuals on this change in future income risk can be analyzed with the second part of the growth rate (26). The impact of the stochastic pollution tax on equilibrium growth depends on the intertemporal elasticity of substitution, as this elasticity decides upon the dominance of income or substitution effects. Any positive stochastic pollution tax T^{s} will reduce the volatility of capital return as can be seen from the variance of capital (24). Actually, environmental taxation can provide a complete insurance against the income risk caused by environmental risk if the stochastic pollution tax takes the form $T_i^{\mathrm{s}} = Ak_i\eta_T^{-\alpha'}$. In this case, the stochastic pollution tax entirely offsets the income uncertainty caused by environmental productivity risk. Hence, the individuals are back in a world with a sure income flow $(\sigma_{k_i}^2 = 0)$ and expected equilibrium growth is given by

$$\varphi = \varepsilon \left(A\eta_T^{\alpha} - T_{k_i}^{\mathrm{d}} - \eta_T(1 + T_{e_i}^{\mathrm{d}}) - \beta\right) \tag{27}$$

and will be analyzed below.

Which attributes characterize the *optimal* pollution tax? Optimal pollution taxation has to adjust equilibrium economic variables to their Pareto-optimal levels. Welfare is determined by the propensity to consume out of wealth, μ, and the abatement ratio, η, as can be seen from Equation (10). Both ratios deviate from their optimal levels. In particular, optimal taxation has to foster individual abatement effort, and to reduce equilibrium consumption. At the same time, equilibrium capital accumulation should not be affected directly, but only through the adjustment of the abatement ratio.

In the model considered here, we find a market failure due to mis-perception with the related consequences for consumption and abatement. Additionally, we find the institutional failure that the idiosyncratic environmental risk is not diversified neither by insurance nor by the government. This institutional failure exists by assumption, but can be justified with the moral hazard argument if the individual productivity risk is unobservable: If the insurance cannot distinguish between low income caused by low individual effort and low income due to a natural disaster, there is no incentive for abatement expenditures.

With respect to efficient pollution taxation, the analysis will be divided into two steps: first, tax schemes are analyzed which correct for the market failure due to mis-perception. Meanwhile, the uncertainty caused by environmental productivity will be regarded as exogenous. In the second step, a tax scheme is analyzed which corrects for the idiosyncratic environmental risk. Of course, this tax scheme can only be realized, if the government is capable to observe the individual realizations of environmental productivity.

5.1. DIRECT POLLUTION TAXATION

There is a negative externality of capital accumulation and a positive externality of abatement activity. Since both are caused by the same mis-perception of the individual influence on pollution (the same δ), a direct pollution tax is suggesting in order to internalize the pollution externality. Just as well as in the deterministic case, it ought to be appropriate to close the gap between individually perceived and socially relevant impact on pollution. Nevertheless, the above stated conditions for optimal pollution taxation cannot be met by a tax, $T^{\mathrm{d}}(P)$, $T^{\mathrm{s}}(P)$, which is levied on pollution directly. In this case, the derivatives of the pollution tax with respect to capital and abatement would result in[15]

$$T^{\mathrm{d}}_{k_i} = (1-\delta)T^{\mathrm{d}'}(\eta_{\mathrm{T}}k_i)^{-1}, \quad T^{\mathrm{s}}_{k_i} = (1-\delta)T^{\mathrm{s}'}(\eta_{\mathrm{T}}k_i)^{-1} \tag{28}$$

$$T^{\mathrm{d}}_{e_i} = -(1-\delta)T^{\mathrm{d}'}(\eta_{\mathrm{T}}e_i)^{-1}, \quad T^{\mathrm{s}}_{e_i} = -(1-\delta)T^{\mathrm{s}'}(\eta_{\mathrm{T}}e_i)^{-1}. \tag{29}$$

Indeed, in a deterministic growth model,[16] this kind of pollution tax – evaluated optimally – leads to the socially optimal steady state (see Smulders and Gradus 1996). The pollution tax has to be determined to equate decentral abatement, η_{T}, as given by (25) and optimal abatement, η, as given by (15). Due to $T^{\mathrm{d}}_{k_i} + \eta_{\mathrm{T}}\, T^{\mathrm{d}}_{e_i} = 0$, the direct pollution taxation implies growth neutrality in the deterministic growth model, as required.

Nevertheless, this result does not extend to the stochastic dynamic setting. Since pollution is constant in any equilibrium, the deterministic as well as the stochastic part of the tax revenue are constant, too. With respect to the stochastic pollution tax, this rules out the possibility of steady state growth. In particular, the growth rate with direct pollution taxation, $T(P)$, becomes

$$\begin{aligned}\varphi_{T(P)} &= \varepsilon(A\eta^{\alpha} - \eta - \beta) \\ &\quad - \frac{\rho}{2}\sigma^2\left((\varepsilon-1)A^2\eta^{-2\alpha'} + 2A\eta^{-\alpha'}\frac{T^{\mathrm{s}}}{k_i} - (1+\varepsilon)\left(\frac{T^{\mathrm{s}}}{k_i}\right)^2\right)\end{aligned} \tag{30}$$

and hence is not independent from the level of capital accumulation. In order to enable steady state growth, the stochastic part of the pollution tax has to increase in capital. Only in this case, the volatility of the pollution tax increases in accordance with the volatility of the income stream induced by uncertain environmental productivity. If instead the stochastic pollution tax is constant, the riskiness of the pollution tax diminishes in relation to the riskiness of income, and steady state growth is impossible.

5.2. LINEAR TAXATION

Since all variables grow with a common rate, the relations between the economic variables remain constant on the steady state growth path. Hence, a simple efficient pollution tax consists in a linear tax on capital income $(\tau^{\mathrm{d}}_k, \tau^{\mathrm{s}}_k)$ combined with a linear subsidy on abatement effort $(\tau_e^{\mathrm{d}}, \tau_e^{\mathrm{s}})$ and with balanced governmental budget

$$T^{\mathrm{d}*} = \tau_k^{\mathrm{d}} k_i + \tau_e^{\mathrm{d}} e_i = 0, \quad T^{\mathrm{s}*} = \tau_k^{\mathrm{s}} k_i + \tau_e^{\mathrm{s}} e_i = 0. \tag{31}$$

This implies growth neutrality due to

$$T_{k_i}^{\mathrm{d}} + \eta_{\mathrm{T}} T_{e_i}^{\mathrm{d}} = 0, \quad T_{k_i}^{\mathrm{s}} + \eta_{\mathrm{T}} T_{e_i}^{\mathrm{s}} = 0. \tag{32}$$

and

$$T^{\mathrm{d}} = 0, \quad T^{\mathrm{s}} = 0. \tag{33}$$

The corresponding equilibrium expected growth rate hence coincides with the Pareto-optimal growth rate (16).

In order to adjust the abatement ratio, the optimal levels of the constant tax rates, τ_k^{d}, τ_e^{d}, τ_k^{s}, τ_e^{s}, become relevant. The balanced governmental budget implies

$$\tau_k^{\mathrm{d}} = -\eta_{\mathrm{T}} \tau_e^{\mathrm{d}}, \quad \tau_k^{\mathrm{s}} = -\eta_{\mathrm{T}} \tau_e^{\mathrm{s}}. \tag{34}$$

The optimal solutions for the tax rates are evident from the equalization of equilibrium abatement, η_{T}, and optimal abatement, η,

$$\tau_k^{\mathrm{d}*} = \delta(\gamma\mu + \alpha A\eta^{\alpha}), \quad \tau_k^{\mathrm{s}*} = -\delta\alpha' A\eta^{-\alpha'} \tag{35}$$

$$\tau_e^{\mathrm{d}*} = -\frac{\delta}{\eta}(\gamma\mu + \alpha A\eta^{\alpha}), \quad \tau_e^{\mathrm{s}*} = \frac{\delta}{\eta}\alpha' A\eta^{-\alpha'}. \tag{36}$$

The deterministic part of the pollution tax accounts for the pollution externalities in consumption as well as in production. The larger the parameter δ, the more influential is the individual mis-perception. Individuals perceive a smaller part of their influence on pollution. Hence, there is more need for internalization, and capital income taxation as well as abatement subsidy rise (in absolute value). Note, that the tax on the stochastic component of capital income in fact is a subsidy $(\tau_k^{\mathrm{s}*} < 0)$. The reason is that households perceive only part of the riskiness of environmental productivity to depend on their individual decisions. The subsidy on the stochastic component of capital returns closes the gap between individually perceived and socially relevant environmental risk. With respect to the subsidy on abatement, the argument reverses: The stochastic part of abatement expenditure is taxed in order to reduce the incorporated uncertainty and to increase the attractiveness of abatement for risk averse individuals.

Moreover, the deterministic part of the pollution tax is affected by environmental risk via the optimal consumption and abatement ratios, η and μ. The impact of uncertainty is ambiguous, as already illustrated in Section 3. In the special case $\alpha = 0$ which neglects the property of environmental quality as a determinant of productivity and focuses on the disutility out of pollution,[17] the impact of uncertainty on the optimal deterministic pollution tax is given by

$$\frac{\partial \tau_k^{\mathrm{d}*}}{\partial \sigma^2} = \delta\gamma \frac{\partial \mu}{\partial \sigma^2} = -\frac{\delta\gamma\rho A^2(1-\varepsilon)}{2(1+\gamma(1-\varepsilon))} < 0, \quad \varepsilon < 1. \tag{37}$$

With an elasticity of substitution, ε, which is below unity, optimal growth and optimal pollution increase in reaction on environmental risk. Hence, there is less need for growth reduction by means of income taxation.

5.3. INSURANCE AGAINST INCOME RISK

Up to now, efficient pollution taxation was defined to correct for the market failure due to mis-perception. Uncertainty associated with pollution was considered arbitrarily. As long as environmental productivity risk is aggregate, or realization is not observable, there is an institutional failure which impedes any insurance and indeed prevents government from risk-pooling. In the following, we will relax this assumption and analyze a tax scheme which pools the income uncertainty associated to pollution by means of income taxation. Hence, welfare can even be enhanced, beyond the situation with arbitrarily given risk.

Whether the environmental risk is idiosyncratic or aggregate, depends on the specific case. Both occurs in reality. Examples may be staff away sick due to environmental reasons or a hurricane which emerges locally. This is the type of risk which is captured by this model with idiosyncratic environmental productivity. In contrast, aggregate environmental risk applies to the whole society: all firms experience the same realization of the productivity shock. An example for aggregate environmental risk is an accident in a nuclear power station.

For idiosyncratic environmental risk, the aggregate income tax revenue out of stochastic pollution taxation is zero. Nevertheless, on the individual level, the volatility of future income streams decreases with a rise in the stochastic pollution tax. This insurance argument of income taxation draws back on Domar and Musgrave (1944) and was further developed by Stiglitz (1969). The consequences of taxation on growth and welfare are studied e.g. by Smith (1996b) or Turnovsky (2000) within stochastic growth models. As already discussed with Equation (27), the stochastic pollution tax which entirely absorbs the uncertainty associated with pollution, is given by

$$\tau_k^{\mathrm{s}} = A\eta^{-\alpha'}, \quad \tau_e^{\mathrm{s}} = 0. \tag{38}$$

If furthermore the deterministic pollution tax rates, $\tau_k^{\mathrm{d}*}$ and $\tau_e^{\mathrm{d}*}$, are set optimally according to Equation (35), expected growth yields the first best level

$$\varphi^* = \varepsilon(A\eta^{*\alpha} - \eta^* - \beta). \tag{39}$$

Whether equilibrium growth increases or decreases due to the insurance against environmental risk again depends on the elasticity of intertemporal substitution which determines the dominance of income or substitution effects.

The corresponding first best abatement ratio is given by

$$1 - \alpha A\eta^{*\alpha-1} = \frac{\gamma}{\eta^*}(\varepsilon\beta + (1-\varepsilon)(A\eta^{*\alpha} - \eta^*)). \tag{40}$$

On the one hand, due to the insurance, the volatility associated with environmental degradation decreases. A risk averse society gets "less afraid of" pollution and the optimal environmental quality level, η^*, decreases. In other words,

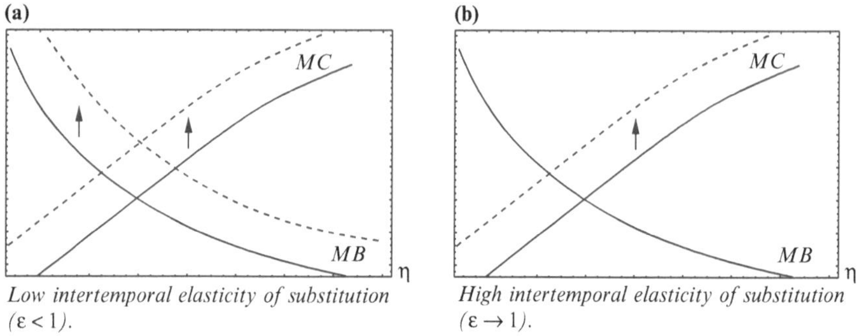

Figure 3. Complete insurance and optimal abatement.

the negative marginal cost of abatement due to the reduction of environmental risk vanishes. Therefore, marginal cost of abatement increases. This is the reason for the upward shift of the marginal cost curve (left hand side in equation 40) in Figure 3. η^* ceteris paribus decreases.[18]

On the other hand, due to the insurance, the volatility associated with capital return diminishes. Hence, the certainty equivalent of capital return increases and in the empirically relevant case, $\varepsilon < 1$, optimal savings decrease. Less capital accumulation immediately leads to an increase in environmental quality. η^* ceteris paribus increases. This is indicated with the upward shift of the marginal benefit curve (right hand side in equation 43) in Figure 3.

The over-all impact on optimal abatement is ambiguous and depends predominantly on the productive capacity of pollution, α, which determines the magnitude of the increase in marginal costs, together with the elasticity of intertemporal substitution, ε, and environmental preferences, γ, which determine the magnitude of the increase in marginal benefits. Hence, the first best pollution level, $\eta^{*\wedge -1}$, can increase or decrease due to the insurance against environmental risk.[19]

The magnitude of the upward shifts of the two curves is predominantly determined by intertemporal substitutability. The higher ε, the slighter is the upward shift of marginal benefits, since the change in optimal savings is smaller. In contrast, marginal cost of abatement is independent from inter-temporal substitutability. The limiting case $\varepsilon \to 1$ is displayed in Figure 3b. If the intertemporal elasticity of substitution is sufficiently high, the insurance induces an unambiguous increase in optimal pollution, η^{*-1}. In this case, the pollution increasing effect of the diminished riskiness associated with environmental productivity exceeds the growth increasing (pollution decreasing) effect of the diminished riskiness in capital return.

6. Conclusion

This paper analyzes efficient pollution taxation within a stochastic dynamic framework. Pollution is an inevitable by-product of production and reduces environmental quality. Moreover, there is an amenity value of environmental quality as well as a productivity enhancing effect. The latter is uncertain due to environmental risk: Natural disasters like hurricanes or floods induce capital depreciation which is not known with certainty in advance. The Pareto-optimal steady state is described: The optimal abatement ratio depends positively on the amenity value and the productivity effect of environmental quality. Optimal abatement activity as well as optimal expected growth are affected in ambiguous way by uncertainty. The optimal adjustment of the savings decision on risk depends on the preferences, particularly on the elasticity of intertemporal substitution. It is shown that the riskiness of the environmental productivity effect may lead to a decrease in optimal environmental quality of a risk averse society, if capital accumulation is increased in order to provide an insurance against future income streams.

The individuals are assumed to perceive only part of their impact on pollution. Hence, there are externalities due to mis-perception, and the dynamic equilibrium is inefficient. It is shown that only the intratemporal decision between consumption and abatement is influenced. The intertemporal savings decision instead is independent from mis-perception.

In order to establish an efficient pollution tax, a general formulation of pollution taxation is introduced and the market equilibrium with arbitrarily given pollution taxation is determined. Two conditions are set up: An efficient pollution tax has to adjust the abatement effort to the optimal level and beyond has to be growth neutral. Steady state growth turns out to be inconsistent with a pollution tax which is levied directly on pollution. The resulting tax payment in this case is characterized by constant volatility, whereas income volatility increases through time. Hence, the relative riskiness changes and inhibits steady state growth. Instead, a linear pollution tax which is levied on capital income and abatement effort, can be used to internalize the externalities.

A first best solution, which additionally corrects for the institutional failure of absence of risk-pooling, can be obtained if the government uses the stochastic capital income tax in order to provide an insurance against environmental risk. With complete insurance, uncertainty vanishes. Nevertheless, optimal intertemporal savings as well as the optimal pollution level react ambiguously on the insurance, depending on intertemporal substitution. Hence, the optimal environmental quality may decrease due to insurance, since risk averse individuals get less afraid of the volatility of environmental risk.

Acknowledgements

I have benefited from discussions at SURED conference in Ascona, in the Research Seminar at the ETH Zürich, as well as from useful comments by the editors and three anonymous referees.

Notes

1. The assumption of a general pollution function $P_i(t) = P(k_i(t), e_i(t))$ with constant and equal elasticities of pollution with respect to capital and abatement would end up with the same results.

2. As a consequence, individuals choose equal growth rates of capital and abatement in any equilibrium. Thus, any equilibrium will be sustainable due to constant pollution, even though pollution will be suboptimally high in equilibrium.
3. Instead of p_i, the stochastic process of productivity could also be defined by dp_i. In order to emphasize that pollution itself is a flow variable, I favor the notation p_i.
4. With the assumption of a Wiener process, I restrain the analysis to marginal shocks. Of course, various types of environmental quality would be characterized better with jumps in the pollution level. Nevertheless, Steger (2005) shows that the implications of marginal shocks (a Wiener process) and jumps (a Poisson process) are qualitatively the same. In order to keep the model simple, I assume environmental uncertainty to be marginal.
5. The results remain qualitatively unchanged for any constant population size N.
6. Nevertheless, a decrease in the volatility of environmental productivity due to lower environmental quality can be analyzed with the same model simply by setting $\alpha' < 0$.
7. Due to intergenerational altruism, the individuals can be interpreted as long-lived dynasties.
8. If an increase in pollution causes a decrease in the volatility of environmental productivity (if $\alpha' < 0$), the certainty equivalent of capital returns has to be positive in order to enable feasible solutions, that is, $-\rho\alpha/\alpha' A\eta^{-\alpha}\sigma^2 < 1$. This condition is equivalent to the requirement that uncertainty should not be too strong.
9. Only for the special case $\alpha = \alpha' = 0$, the solution is suggesting: optimal abatement in this case is given by $\eta = (\varepsilon\beta + (1-\varepsilon)A - \rho/2(1-\varepsilon)A^2\sigma^2)/(1/\gamma + 1 - \varepsilon)$ and optimal consumption results in $\mu = \gamma\eta$. This case is discussed with detail in Soretz (2003).
10. A detailed discussion is relocated to the appendix. I am indebted to an anonymous referee for this solution procedure.
11. For the distinction between risk premia (which depend on the intertemporal elasticity of substitution) and the motive for precautionary savings (which is based on risk aversion), see e.g. Kimball (1990), Weil (1993) or Gollier et al. (2000).
12. As already shown by Smith (1996a), positive consumption (feasibility) is neither necessary nor sufficient for the transversality condition to be satisfied. Instead, both conditions have to be verified separately. Nevertheless, Smith (1996a) proves that the feasibility condition as well as the transversality condition are met for empirically relevant parameterization: A relatively low intertemporal elasticity of substitution, $\varepsilon \leqslant 1$ is a sufficient condition for feasibility and a relatively high degree of risk aversion, $\rho \geqslant 1$, automatically satisfies the transversality condition. Subsequently, only parameter settings will be considered which satisfy the feasibility as well as the transversality condition.
13. See appendix for the derivation.
14. This outcome corresponds to the well-known growth neutrality of consumption tax.
15. In order to maintain consistency within individual optimization, I assume that individuals suppose the tax to depend on *perceived* pollution. Nevertheless, the argument is the same if the tax directly depends on "true" pollution.
16. The corresponding deterministic growth model is rapidly described by setting $\sigma^2 = 0$.
17. In this setting the Pareto-optimal consumption and abatement ratios result in $\mu = \frac{1}{1+\gamma(1-\varepsilon)}(\varepsilon\beta + (1-\varepsilon)A(1-\rho A\sigma^2/2))$ and $\eta = \gamma\mu$.
18. Gaube (2005) develops a related outcome in a deterministic framework: He shows that environmental quality may be higher in a second best situation with distortionary taxation than in a first best optimum with lump-sum taxes.
19. The case of decreasing marginal cost is not depicted with detail. Nevertheless, the results apply to both cases.

References

Ayong Le Kama, A. and K. Schubert (2004), 'Growth, Environment and Uncertain Future', *Environmental and Resource Economics* **28**, 31–53.

Baranzini, A. and F. Bourguignon (1995), 'Is Sustainanble Growth Optimal?', *International Tax and Public Finance* **2**, 341–356.

Beltratti, A., G. Chichilnisky and G. M. Heal (1998). 'Uncertain Future Preferences and Conservation', in: G. Chichilnisky, G. M. Heal and A. Vercelli, eds., *Sustainability: Dynamics and Uncertainty* (pp. 257–275). Dordrecht: Kluwer Academic Publishers.

Bovenberg, A. L. and S. Smulders (1997), 'Environmental Quality and Pollution-Augmenting Technological Change in a Two-Sector Endogenous Growth Model', *Journal of Public Economics* **57**(3), 153–179.

Byrne, M. M. (1997), 'Is Growth a Dirty Word? Pollution, Abatement and Endogenous Growth', *Journal of Development Economics*, **54**, 261–284.

Chichilnisky, G. and G. M. Heal (1998), 'Financial Markets for Unknown Risks', in G. Chichilnisky, G. M. Heal and A. Vercelli, eds., *Sustainability: Dynamics and Uncertainty* (pp. 277–294). Dordrecht: Kluwer Academic Publishers.

Clemens, C. and S. Soretz (1997), 'Macroeconomic Effects of Income Taxation in a Model of Stochastic Growth', *Finanzarchiv, N. F.*, **54**(4), 471–493.

Clemens, C. and S. Soretz (2004), 'Optimal Fiscal Policy, Uncertainty, and Growth', *Journal of Macroeconomics* **26**, 679–697.

Corsetti, G. (1997), 'A Portfolio Approach to Endogenous Growth: Equilibrium and Optimal Policy', *Journal of Economic Dynamics and Control* **21**, 1627–1644.

Domar, E. D. and R. A. Musgrave (1944), 'Proportional Income Taxation and Risk-Taking', *The Quarterly Journal of Economics* **59**, 388–422.

Duffie, D. and L. G. Epstein (1992), 'Stochastic Differential Utility', *Econometrica* **60**, 353–394.

Eaton, J. (1981), 'Fiscal Policy, Inflation and the Accumulation of Risky Capital', *Review of Economic Studies* **48**, 435–445.

Edwards, J. H. Y. (1990), 'Congestion Function Specification and the 'Publicness' of Local Public Goods', *Journal of Urban Economics* **27**, 80–96.

Epstein, L. G. and S. E. Zin (1989), 'Substitution, Risk Aversion, and the Temporal Behaviour of Consumption and Asset Returns: A Theoretical Framework', *Econometrica* **57**, 937–969.

Epstein, L. G. and S. E. Zin (1991), 'Substitution, Risk Aversion, and the Temporal Behaviour of Consumption and Asset Returns: An Empirical Analysis', *Journal of Political Economy* **99**(2), 263–286.

Eriksson, C. (2004), 'Can Green Consumerism Replace Environmental Regulation?—A Differentiated–Products Example', *Resource and Energy Economics* **26**, 281–293.

Fernandez, L. (2005), 'A Diversified Portfolio: Joint Management of Non-renewable and Renewable Resource Offshore', *Resource and Energy Economics* **27**, 65–82.

Fisher, W. H. and S. J. Turnovsky (1998), 'Public Investment, Congestion, and Private Capital Accumulation', *The Economic Journal* **108**, 1 399–413.

Gaube, T. (2005), 'Second-Best Pollution Taxation and Environmental Quality', *Frontiers of Economic Analysis & Policy* **1**(1), 1–16.

Glomm, G. and B. Ravikumar (1994), 'Public Investment in Infrastructure in a Simple Growth Model', *Journal of Economic Dynamics and Control* **18**, 1173–1187.

Gollier, C., B. Jullien and N. Treich (2000), 'Scientific Progress and Irreversibility: An Economic Interpretation of the "Precautionary Principle", *Journal of Public Economics* **75**, 229–253.

Gradus, R. and S. Smulders (1993), 'The Trade-off Between Environmental Care and Long-term Growth—Pollution in Three Prototype Growth Models', *Journal of Economics* **58**, 25–51.

Hall, R. E. (1988), 'Intertemporal Substitution in Consumption', *Journal of Political Economy* **96**(2), 339–357.

Hartman, R. and O.-S. Kwon (2005), 'Sustainable growth and the environmental Kuznets curve', *Journal of Economic Dynamics and Control* **29**, 1701–1736.

Jones, L. E. and R. E. Manuelli (1995), 'A Positive Model of Growth and Pollution Controls', *NBER Working Paper Series 5205*, National Bureau of Economic Research, Cambridge, MA.

Keller, K., B. M. Bolker and D. F. Bradford (2004), 'Climate Thresholds and Optimal Economic Growth', *Journal of Environmental Economics and Management* **48**, 723–741.

Kimball, M. S. (1990), 'Precautionary Saving in the Small and in the Large', *Econometrica* **58**(1), 53–73.
Lafforgue, G. (2005), 'Uncertainty and Amenity Values in Renewable Resource Economics', *Environmental and Resource Economics* **31**, 369–383.
Ligthart, J. E. and F. van der Ploeg (1994), 'Pollution, the Cost of Public Funds and Endogenous Growth', *Economics Letters* **46**, 351–361.
Obstfeld, M. (1994), 'Risk-Taking, Global Diversification, and Growth', *The American Economic Review* **84**, 1310–1329.
Smith, W. T. (1996), 'Feasibility and Transversality Conditions for Models of Portfolio Choice with Non-Expected Utility in Continuous Time', *Economics Letters* **53**, 123–131.
Smith, W. T. (1996b), 'Taxes, Uncertainty, and Long-term Growth', *European Economic Review* **40**(8), 1647–1664.
Smulders, S. and R. Gradus (1996), 'Pollution Abatement and Long-Term Growth', *European Journal of Political Economy* **12**, 505–532.
Soretz, S. (2003), 'Stochastic Pollution and Environmental Care in an Endogenous Growth Model', *The Manchester School* **71**, 448–469.
Soretz, S. (2004), 'Stochastic Environmental Policy, Risk-Taking, and Growth: Discretion versus Commitment', *International Journal of Global Environmental Issues* **4**, 53–71.
Steger, T. (2005), 'Stochastic Growth under Wiener and Poisson Uncertainty', *Economics Letters* **86**, 311–316.
Stiglitz, J. E. (1969), 'The Effects of Income, Wealth, and Capital Gains Taxation on Risk-Taking', *The Quarterly Journal of Economics* **83**, 263–283.
Stokey, N. L. (1998), 'Are There Limits to Growth?', *International Economic Review* **39**, 1–31.
Svensson, L. E. O. (1989), 'Portfolio Choice with Non-expected Utility in Continuous Time', *Economics Letters* **30**, 313–317.
Turnovsky, S. J. (1993), 'Macroeconomic Policies, Growth, and Welfare in a Stochastic Economy', *International Economic Review* **34**, 953–981.
Turnovsky, S. J. (1995), *Methods of Macroeconomic Dynamics*, Cambridge MA.: MIT Press.
Turnovsky, S. J. (1995), 'Optimal Tax Policy in a Stochastically Growing Economy', *The Japanese Economic Review* **46**, 125–147.
Turnovsky, S. J. (1999), 'Productive Government Expenditure in a Stochastically Growing Economy', *Macroeconomic Dynamics* **3**, 544–570.
Turnovsky, S. J. (2000), 'Fiscal Policy, Elastic Labor Supply, and Endogenous Growth', *Journal of Monetary Economics* **45**, 185–210.
Weil, P. (1990), 'Non-expected Utility in Macroeconomics', *The Quarterly Journal of Economics* **105**, 29–42.
Weil, P. (1993), 'Precautionary Savings and the Permanent Income Hypothesis', *Review of Economic Studies* **60**, 367–383.

Appendix

Determination of the social optimum

Existence and unicity of the social optimum are derived by the analysis of marginal cost, MC, and marginal benefit, MB, of abatement effort. According to condition (15), marginal cost and marginal benefit are given by

$$\mathrm{MC}(\eta) = 1 - \alpha A \eta^{\alpha-1}\left(1 + \rho\frac{\alpha'}{\alpha} A \eta^{-2\alpha'-\alpha}\sigma^2\right)$$

$$\mathrm{MB}(\eta) = \frac{\gamma}{\eta}\left(\varepsilon\beta + (1-\varepsilon)\left(A\eta^{\alpha} - \eta - \frac{\rho}{2}A^2\eta^{-2\alpha'}\sigma^2\right)\right). \qquad \text{(A.1)}$$

The thread is as follows: MB is unambiguously decreasing in abatement if the empirically relevant case $\varepsilon < 1$ applies. The slope of MC can be either positive or negative. Therefore, it is shown that MC does not decrease faster than MB. Moreover, MB is shown to be greater than MC for small enough values of η and less than MB at the upper bound of feasible η. These arguments together give existence and unicity of the optimal abatement ratio.

The slope of marginal benefit evolves to

$$\begin{aligned} \mathrm{MB}' \equiv \frac{\partial \mathrm{MB}}{\partial \eta} &= -\frac{\gamma\mu}{\eta^2} - \frac{\gamma}{\eta}(1-\varepsilon)\left(1 - \alpha A \eta^{\alpha-1}\left(1 + \rho\frac{\alpha'}{\alpha}A\eta^{-2\alpha'-\alpha}\sigma^2\right)\right) \\ &= -\frac{\mathrm{MB}}{\eta} - \frac{\gamma}{\eta}(1-\varepsilon)\mathrm{MC}. \end{aligned} \tag{A.2}$$

In the neighborhood of the intersection between marginal benefit and marginal cost, MB can be approximated by MC. Hence, the slope of MB in the intersection with MC is given by

$$\mathrm{MB}' = -\frac{\mathrm{MC}}{\eta}(1 + \gamma(1-\varepsilon)). \tag{A.3}$$

In the empirically relevant case, $\varepsilon < 1$, marginal benefit decreases with a rise in abatement, $\mathrm{MB}' < 0$.

Nevertheless, the slope of MC is ambiguous, as can be seen with

$$\mathrm{MC}' \equiv \frac{\partial \mathrm{MC}}{\partial \eta} = -\alpha(\alpha-1)A\eta^{\alpha-2} + \alpha'(2\alpha'+1)\rho A^2\eta^{-2\alpha'-2}\sigma^2 \gtrless 0 \tag{A.4}$$

$$\Longleftrightarrow \eta \lesseqgtr \left(\frac{\alpha(\alpha-1)}{\alpha'(2\alpha+1)\rho A\sigma^2}\right)^{-\frac{1}{2\alpha'+\alpha}}. \tag{A.5}$$

For small values of abatement, MC increases, for large values, it decreases. Only for $\alpha \leqslant 1$, the outcome is clear cut: (A.4) shows that $\mathrm{MC}' > 0\ \forall \eta > 0$. In the case $\alpha > 1$ it is not possible to exclude the maximum of MC from the range feasible abatement effort: An abatement ratio is feasible if it is positive and below expected capital productivity, since abatement effort cannot exceed average income. Hence, $0 < \eta < A^{\frac{1}{1+\alpha}}$ describes the range of feasible abatement effort, and the maximum given in Equation (A.5) may well be situated within this interval.

Hence, to prove existence and unicity of optimal abatement, it will be shown that around any intersection of MB and MC, the slope of MC is greater than that of MB, that is, $\mathrm{MC}' > \mathrm{MB}'$. To show this relation, the second order necessary condition with respect to e has to be determined

$$\begin{aligned} \frac{\partial^2 \mathcal{B}}{\partial e^2} &\Longrightarrow \gamma(\gamma(1-1/\varepsilon)-1)\mu\eta^{-2} + \alpha(\alpha-1)A\eta^{-\alpha-2} \\ &\quad -\rho A^2\alpha'(2\alpha'+1)\eta^{-2\alpha'-2}\sigma^2 \overset{!}{<} 0 \\ &\Longleftrightarrow \gamma(\gamma(1-1/\varepsilon)-1)\frac{\eta}{\gamma}\mathrm{MC}\eta^{-2} - \mathrm{MC}' < 0. \end{aligned} \tag{A.6}$$

From Equation (A.3) follows immediately $\mathrm{MC} = \mathrm{MB}'\eta/(1 + \gamma(1-\varepsilon))$ which is valid in the neighborhood of optimal abatement, and therefore

$$\frac{\gamma(1-1/\varepsilon)-1}{1+\gamma(1-\varepsilon)}\mathrm{MB}' + \mathrm{MC}' > 0 \Longleftrightarrow \frac{1+1/\varepsilon\gamma(1-\varepsilon)}{1+\gamma(1-\varepsilon)}\mathrm{MB}' < \mathrm{MC}'. \tag{A.7}$$

The fraction in Equation (A.7) is greater than one as long as ε is below unity. Hence, the second order condition (A.7) is sufficient for MB' to be less than MC'.

The remaining points are to verify that *MB* is greater (smaller) than MC for small (large) values of η. At the upper bound of feasible abatement effort, $\bar{\eta} \equiv A^{\frac{1}{1+\alpha}}$, expected growth as well as consumption vanish, $\varphi(\bar{\eta}) = -\mu(\bar{\eta}) = 0$, and therefore

$$\mathrm{MB}(\bar{\eta}) = \frac{\gamma}{\bar{\eta}}\mu(\bar{\eta}) = 0. \tag{A.8}$$

Due to $\mathrm{MC}(\bar{\eta}) > 0$, marginal cost is greater than marginal benefit at this upper bound of abatement activity.

With abatement activity sufficiently low, MC gets negative

$$\lim_{\eta\to 0} \mathrm{MC} = \lim_{\eta\to 0} 1 - \alpha A\eta^{\alpha-1}\left(1 + \rho\frac{\alpha'}{\alpha}A\eta^{-2\alpha'-\alpha}\sigma^2\right) = -\infty. \tag{A.9}$$

From MB $> 0\ \forall\eta$, it is obvious that marginal benefit is positive. Therefore, marginal benefit is greater than marginal cost at the lower bound of abatement activity. Consequently, there has to be exactly one intersection between marginal cost and marginal benefit, which is situated in the feasible range and indicates socially optimal abatement effort. The argument is summarized in Figure 1 in Section 3.

Market equilibrium

Dynamic market equilibrium is determined by the maximization of the stochastic Bellman equation

$$\begin{aligned}\mathcal{B} = {} & \frac{1-\rho}{1-1/\varepsilon}(c_i P_{\mathrm{p}}^{-\gamma})^{1-1/\varepsilon} - \beta G((1-\rho)J_i(k_i)) \\ & + (1-\rho)G'((1-\rho)J_i(k_i))\left(J_i'(k_i)\frac{\mathrm{E}[dk_i]}{dt} + \frac{1}{2}J_i''(k_i)\sigma_{k_i}^2\right)\end{aligned} \tag{A.10}$$

with the variance of individual capital, $\sigma_{k_i}^2 = A^2k_i^2P_{\mathrm{p}}^{2\alpha'}\sigma^2$. The derivation of the Bellman equation with respect to consumption remains unchanged as in (8) and together with the conjecture of constant and equal consumption and abatement ratios, μ and η, yields the same guess for the value function (10).

The first order conditions with respect to abatement effort and capital accumulation now have to account for perceived pollution:

$$\gamma(1-\delta)(c_iP_{\mathrm{p}}^{-\gamma})^{1-1/\varepsilon} + G_i'\left(J_i'(\alpha(1-\delta)Ak_iP_{\mathrm{p}}^{-\alpha} - e_i) + \frac{1}{2}J_i''e_i\frac{\partial\sigma_{k_i}^2}{\partial e_i}\right) \stackrel{!}{=} 0 \tag{A.11}$$

$$\begin{aligned}& -\gamma(1-\delta)(cP_{\mathrm{p}}^{-\gamma})^{1-1/\varepsilon}k_i^{-1} + (1-\rho)G_i''J_i'\left(J_i'\frac{\mathrm{E}[dk_i]}{dt} + \frac{1}{2}J_i''\sigma_{ki}^2\right) \\ & + G_i'\left(J_i'(AP_{\mathrm{p}}^{-\alpha}(1-\alpha(1-\delta)) - \beta) + J_i''\left(\frac{\mathrm{E}[dk_i]}{dt} + \frac{1}{2}\frac{\partial\sigma_{k_i}^2}{\partial k_i}\right) + \frac{1}{2}J_i'''\sigma_{k_i}^2\right) \stackrel{!}{=} 0.\end{aligned} \tag{A.12}$$

Transformation of these conditions follows the same procedure as described above with the Pareto-Optimum. Only the partial derivatives of perceived pollution change.

With respect to the dynamic market equilibrium with pollution tax, the Bellman equation remains unchanged as given in Equation (A.10) for the dynamic market equilibrium. Again, the derivative with respect to consumption is given in (8) and together with the conjecture of constant and equal growth rates (9) leads to the same guess of the value function as derived in (10).

The derivatives with respect to individual abatement expenditures and capital accumulation result in

$$\begin{aligned}& \gamma(1-\delta)(c_iP_{\mathrm{p}}^{-\gamma})^{1-1/\varepsilon} \\ & \quad + G_i'\left(J_i'(\alpha(1-\delta)Ak_iP_{\mathrm{p}}^{-\alpha} - e_i(1+T_{e_i}^{\mathrm{d}})) + \frac{1}{2}J_i''e_i\frac{\partial\sigma_{k_i}^2}{\partial e_i}\right) \stackrel{!}{=} 0\end{aligned} \tag{A.13}$$

$$\begin{aligned}& -\gamma(1-\delta)(cP_{\mathrm{p}}^{-\gamma})^{1-1/\varepsilon}k^{-1} + (1-\rho)G_i''J_i'\left(J_i'\frac{\mathrm{E}[dk_i]}{dt} + \frac{1}{2}J_i''\sigma_{k_i}^2\right) \\ & + G_i'\left(J_i'(AP_{\mathrm{p}}^{-\alpha}(1-\alpha(1-\delta)) - T_{k_i}^{\mathrm{d}} - \beta) + J_i''\left(\frac{\mathrm{E}[dk_i]}{dt} + \frac{1}{2}\frac{\partial\sigma_{k_i}^2}{\partial k_i}\right) + \frac{1}{2}J_i'''\sigma_{k_i}^2\right) \stackrel{!}{=} 0.\end{aligned} \tag{A.14}$$

with the respective derivatives of the variance of capital given by

$$\frac{\partial \sigma_{k_i}^2}{\partial e_i} = -2\sigma^2 \eta^{-1} (A\eta^{-\alpha'} k_i - T^{\mathrm{s}})(\alpha'(1-\delta)A\eta^{-\alpha'} - \eta T_{e_i}^{\mathrm{s}}) \tag{A.15}$$

$$\frac{\partial \sigma_{k_i}^2}{\partial k_i} = 2\sigma^2 (A\eta^{-\alpha'} k_i - T^{\mathrm{s}})((1+\alpha'(1-\delta))A\eta^{-\alpha'} - T_{k_i}^{\mathrm{s}}). \tag{A.16}$$

7. A New-Growth Perspective on Non-Renewable Resources

CHRISTIAN GROTH

Department of Economics and EPRU, University of Copenhagen. Studiestraede 6, DK-1455, Copenhagen K, Denmark (e-mail: chr.groth@econ.ku.dk)*

Key words: Endogenous growth, innovation, non-renewable resources, knife-edge conditions, robustness, limits to growth.

1. Introduction

The aim of this article is to review issues related to the incorporation of scarce natural resources in the theory of economic growth and development. More specifically, we shall concentrate on the role of non-renewable resources. A *non-renewable resource* is a natural resource the amount of which on earth is finite and which has no natural regeneration process (at least not within a relevant time scale). Hence, the stock of a non-renewable resource is depletable. Fossil fuels as well as many non-energy minerals are examples. A *renewable resource* is also available only in limited supply, but its stock is replenished by a natural regeneration process. Hence, if the stock of a renewable resource is not over-exploited, it can be sustained in a more or less constant amount. Fertile soil, fish in the sea, and environmental qualities (clean air etc.) would be examples. In this article the focus is on the specific features of *non-renewable* resources in relation to the feasibility of sustained economic growth.

The old Malthusian and Ricardian views were that scarce natural resources tend to cause diminishing returns to inputs of capital and labour taken together and thereby economic stagnation in the long run. Malthus and Ricardo had primarily land in mind. But what if also non-renewable, hence exhaustible, resources are essential inputs in production? Then the long-run prospect may be worse than stagnation according to the dire predictions of the Club of Rome set forth in the "Limits to growth" report by Meadows et al. (1972).[1] The worldwide oil crisis of the mid-1970s fuelled the interest in this topic.[2] Prominent economists like Solow (1974a, 1974b), Stiglitz (1974a, 1974b), Dasgupta and Heal (1974), and others took these challenges as an occasion for in-depth studies of the macroeconomics of non-renewable resources, including the big questions about sustainable

L. Bretschger and S. Smulders (eds.), Sustainable Resource Use and Economic Dynamics, 127–163.

development, defined as non-decreasing standard of living, or even sustained economic growth. Many issues were clarified, but since the big questions were essentially embedded in a framework with exogenous future technology (hence, unforeseeable), definitive answers could not be given. Although growth has not been hindered by resource shortages in the past, it is another thing whether this can continue in the future.

Beginning with the contributions by Paul Romer (1986, 1987, 1990) and Robert Lucas (1988) there has been, since the late 1980s, a surge of so-called new growth theory or endogenous growth theory. Characteristic traits of this theoretical development are: (1) the focus on conditions that allow endogenous sustained productivity growth; and (2) the systematic incorporation of "ideas" (with their distinctive properties compared with other economic goods) into dynamic general equilibrium models with imperfect competition. In particular there have been great advances in the understanding of technological change. In this article we shall therefore ask:

> What light does new growth theory throw on the limits-to-growth question?

Since there have been several controversies (e.g. about scale effects of different kinds or non-robustness due to knife-edge assumptions) within new growth theory, we add the additional question:

> Does the existence of non-renewable resources have anything to say in relation to the controversies within new growth theory?

It turns out that a key distinction (which has not always received the requisite attention) is that between models where essential non-renewable resources are growth-essential and models where they are not. A non-renewable resource is called *growth-essential* if it is a necessary input to the growth-generating sector(s), the "growth engine", in the economy. It can be so either directly or indirectly by being essential for the manufacturing sector which then delivers necessary input to the "growth engine", usually an R&D or educational sector. Indeed, we shall see that whether non-renewable resources are growth-essential or not has non-trivial implications for the limits-to-growth question.

The remainder of the chapter discusses these issues within a unified framework. Section 2 gives an overview of new growth theory. Section 3 portrays the wave of natural resource economics of the 1970s. In Section 4 a simple one-sector growth model with endogenous technical change is introduced. Section 5 considers different approaches to two-sector models with non-renewable resources and endogenous technical change. The analysis lays bare the key role of the distinction between resources that are growth-essential and resources that are not. Section 6 debates the implications and briefly comments on other research directions, whereas Section 7 summarizes.[3]

2. New Growth Theory

Before considering the integration of non-renewable resources into new growth theory, let us recapitulate the key ingredients of new growth theory as such. The surge of new growth theory or endogenous growth theory began with Romer (1986, 1987, 1990) and Lucas (1988). The term *endogenous growth* refers to models where sustained positive growth in output per capita is driven by some internal mechanism (in contrast to exogenous technology growth).[4]

It is common to divide the endogenous growth literature into two broad classes: accumulation-based models and innovation-based models. The first class of models is based on the idea that the combination of physical and human capital accumulation may be enough to sustain long-run productivity growth. These contributions include the human capital model by Lucas (1988) and the "AK model" by Rebelo (1991). The second class of models, which is more central to our theme here, attempts to explain how technological change comes about and how it shapes economic growth. Technological progress is seen as evolving from purposeful decisions by firms in search for monopoly profits on innovations. An important ingredient in this approach is therefore an attempt at incorporating other market structures than perfect competition into a macroeconomic framework.

Within the class of innovation-based growth models we shall make a distinction between "first-generation" models and "second-generation" models. The first-generation models concentrated on *either* horizontal *or* vertical innovations. The second-generation models integrated these two one-sided lines of attack.

2.1. FIRST-GENERATION MODELS

The first-generation innovation-based growth models have their origin in Romer (1987, 1990), where growth is driven by specialization and increasing division of labour. That is, the focus is on *horizontal innovations:* the invention of new intermediate or final goods gives rise to new branches of trade. The invention of microprocessors is an example. Shortly after the Romer papers came out, Grossman and Helpman (1991, Chapter 4) and Aghion and Howitt (1992) proposed theories in which growth is driven by *vertical innovations*. This strand of endogenous growth theory concentrates on the invention of better qualities of existing products and better production methods that make previous qualities and methods obsolete; improvement in the performance of microprocessors provides an example. The two kinds of models are often called *increasing variety* models versus *increasing quality* models (or *quality ladder* models), respectively.

For both kinds of models the typical set-up is a two-sector framework. There is a manufacturing sector whose output is used for consumption as well as investment in capital of different varieties *or* new qualities (making the previous quality obsolete). The other sector is the "innovative sector". In this sector two activities take place. Firstly, there is R&D activity leading to new capital-good varieties or new capital-good qualities. Secondly, once the technical design (blueprint) of a new variety or quality has been invented, the inventor starts supplying capital goods in the new form, protected by a patent or some kind of secrecy. The key feature behind the generation of sustained per capita growth in both the increasing variety models and the increasing quality models is the assumption of non-diminishing returns to the producible direct or indirect input(s) in the *growth-engine*, i.e. the sector or sectors that "drive growth".[5] Usually the models are structured such that the innovative sector only uses (non-producible) labour as a direct input and therefore, by itself, constitutes the growth-engine. But the productivity of this labour input depends positively on society's accumulated technical knowledge, hence this stock of knowledge can be seen as a produced indirect input.[6] Then non-diminishing returns to knowledge are needed to generate positive per capita growth. In practice exactly constant returns to knowledge (at least asymptotically) are assumed. This is because with increasing returns, growth would explode (see below).

Adding a description of the market structure and households' preferences, the model can be solved. When certain parameter restrictions are satisfied two kinds of results stand out:

- Growth is *fully endogenous*[7] in the sense that the long-run growth rate in output per capita is positive without the support of growth in any exogenous factor; the key to this is the assumption of constant returns to the producible input(s) in the growth engine.
- Via influencing incentives, *policy can affect growth* not only temporarily (i.e. during the transition to a new steady growth path), but also permanently (by affecting the slope of the steady growth path). This is in contrast to the traditional neoclassical growth models, like the Solow model or the Ramsey model, where economic policy (e.g. an investment subsidy) can have only a *level effect* in the long run.

An unwelcome implication of the models is the *scale effect on growth*. Indeed, the models imply the counterfactual predictions: (a) the larger the population is, ceteris paribus, the higher is the long-run per capita growth rate; and (b) sustained growth in population should be associated with a forever rising per capita growth rate. In fact, because of this scale effect the first-generation models simply ignore population growth and assume a constant labour force.

The scale effect is linked to the fact that technical knowledge, by which we mean a set of instructions or recipes about how to combine various inputs to obtain a specific output, is very different from ordinary economic goods in that it is a *non-rival* good. The use of knowledge by one agent does not in itself limit the simultaneous use of the same piece of knowledge by another agent or by many people. In this respect knowledge is dissimilar to human capital, which is embodied in an individual and therefore a rival good. The non-rival character of knowledge implies that output per capita depends on the *total* stock of ideas, not on the stock per person. A larger population breeds more ideas, leading to higher productivity. In the fully endogenous growth models, due to the (knife-edge) assumption of constant returns to knowledge, this takes the extreme form of a scale effect not just on the level of output per capita, but on its growth rate.

The fact that technical knowledge is a non-rival good and only partially excludable (by patents, concealment etc.) makes it a very peculiar good which gives rise to market failures of many kinds. Thus, government intervention becomes an important ingredient in new growth theory.

2.2. THE JONES CRITIQUE AND SEMI-ENDOGENOUS GROWTH

In two important papers, Charles Jones (1995a, 1995b) raised serious concerns about the predictions that not only levels, but also the long-run growth rate, are affected by economic policy and by scale. Jones claimed that: (1) both predictions are rejected by time-series evidence for the industrialized world; (2) both predictions are theoretically non-robust (i.e. they are very sensitive to small changes in parameter values).

The empirical point is supported by, e.g. Evans (1996) and Romero-Avila (2006), although challenged by Li (2002b). As to the theoretical point, let us take Romer's increasing variety model as an example.[8] Consider the aggregate invention production function:

$$\dot{A}(t) \equiv \frac{dA(t)}{dt} = \mu A(t)^{\varphi} L_A(t), \quad \mu > 0, \ \varphi \leqslant 1, \tag{1}$$

where $A(t)$ is the number of existing different capital-good varieties at time t and $L_A(t)$ is research labour, which leads to the invention of new capital-good varieties. The productivity of research labour depends, for $\varphi \neq 0$, on the stock of existing knowledge, which is assumed proportional to $A(t)$. The productivity of labour in manufacturing is similarly assumed proportional to $A(t)$ so that manufacturing output is $Y(t) = F(K(t), A(t)L_Y(t))$, where $K(t)$ and $L_Y(t)$ are inputs of physical capital and labour, respectively, and the production function F is homogeneous of degree one. So far Romer and Jones agree. Their disagreement concerns the likely size of the parameter φ, i.e. the elasticity of research productivity with respect to the level of technical knowledge. In the Romer model, this parameter is (arbitrarily) made equal to one. It may be argued, however, that φ could easily be negative (the "fishing out" case, "the easiest ideas are found first"). Even if one assumes $\varphi > 0$ (i.e. the case where the subsequent steps in knowledge accumulation requires less and less research labour), there is neither theoretical nor empirical reason to expect $\varphi = 1$. The standard "replication argument" for constant returns with respect to the complete set of *rival* inputs is not usable. Even worse, $\varphi = 1$ is a *knife-edge* case. If φ is slightly above 1, then explosive growth arises – and does so in a very dramatic sense: infinite output in finite time. This simple mathematical point is made in Solow (1994). In the numerical example he calculates, the Big Bang – the end of scarcity – is only 200 years ahead! This seems too good to be true.[9]

On the other hand, with φ slightly less than 1, productivity growth peters out, unless assisted by growth in population, an exogenous factor. To see this, let population (= labour force) be $L(t) = L_Y(t) + L_A(t) = L_0 e^{nt}$, where $n \geqslant 0$ is a constant. For any positive variable x, let $g_x \equiv \dot{x}/x$ (the growth rate of x). Then, deriving from equation (1) an expression for $\dot{g}_A/g_A$, we find that in a steady state (i.e. when $\dot{g}_A = \dot{g}_K = \dot{g}_Y = 0$),

$$g_A = \frac{n}{1-\varphi} = g_y, \qquad (2)$$

where y is output per capita ($\equiv Y/L$).[10] There are a number of observations to be made on this result. First, the unwelcome scale effect on growth has disappeared. Second, as indicated by equation (1), a positive scale effect on the *level* of y remains. This is also what we should expect. In view of the non-rival character of knowledge, the per capita cost of creating new knowledge is lower in a larger (closed) society than in a smaller one.[11] Empirically, the "very-long run" history of population and per capita income of different regions of the world gives evidence in favour of scale effects on levels (Kremer 1993). Econometric evidence is provided by, e.g. Alcalá and Ciccone (2004). Third, scale effects on levels also explain why the rate of productivity growth should be an increasing function of the rate of population growth, as implied by equation (2). In view of cross-border technology diffusion, this trait should not be seen as a prediction about individual countries in an internationalized world, but rather as pertaining to larger regions, perhaps the global economy. Finally, unless policy can affect φ or n,[12] long-run growth is independent of policy, as in the old neoclassical story. Of course, "independence of policy" should not be interpreted as excluding that the general social, political, and legal environments can be *barriers* to growth or that, via influencing incentives, policy can affect the long-run *level* of y.

The case $\varphi < 1$ constitutes an example of *semi-endogenous growth*. We say there is semi-endogenous growth when (1) per capita growth is driven by some internal mechanism (as distinct from exogenous technology growth), but (2) sustained per capita growth requires support in the form of growth in some exogenous factor. In innovation-based growth theory, this factor is typically population size. In Jones (1995b), equation (1) takes the extended form, $\dot{A} = \mu A^{\varphi} L_A^{\lambda}$, $0 < \lambda \leqslant 1$, where $1 - \lambda$ represents a likely congestion externality of simultaneous research (duplication of effort); but this externality is not crucial for the discussion here.[13] As we have defined the first-generation models of endogenous growth, the Jones (1995b) model also belongs to this group, being a modified Romer-style increasing-variety growth model. Indeed, whether an analysis concentrates on the robust case $\varphi < 1$ or the non-robust (but analytically much simpler) case $\varphi = 1$, is in our terminology not decisive for what generation the applied model framework belongs to. A further terminological remark is perhaps warranted. Speaking of "fully endogenous" versus "semi-endogenous" growth may give the impression that the first term refers to something going deeper than the second; nothing of that sort should be implied.

2.3. SECOND-GENERATION MODELS

The Jones-critique provoked numerous answers and fruitful new developments. These include different ways of combining the horizontal and the vertical innovation approach (Young 1998, Peretto 1998, Aghion and Howitt 1998, Chap. 12, Dinopoulos and Thompson 1998, Howitt 1999, and Peretto and Smulders 2002).[14] On the one hand these models succeeded in reconciling policy-dependent long-run growth with the absence of a scale effect on growth and thereby the absence of accelerating growth as soon as population growth is present. On the other hand, as maintained by Jones (1999), Li (2000), and Li (2002a), this reconciliation relies on *several* questionable knife-edge conditions; a generic model with innovations along two dimensions tends to have policy-invariant long-run growth, as long as population growth is exogenous, and tends to feature semi-endogenous growth, not fully endogenous growth.[15]

What do these developments within growth theory have to say about the role of natural resources for sustainable development and the role of technological change for overcoming the finiteness of natural resources? In the wake of the first-generation endogenous growth models appeared a series of papers considering the relationship between growth and environmental problems (Brock and Taylor 2005 and Fullerton and Kim 2006 depict the state of the art). Much of this literature does not take the specifics of *non-renewable* resources into account. There has also, however, some work been done on the relationship between endogenous growth and non-renewable resources (Jones and Manuelli 1997,[16] Aghion and Howitt 1998, chapter 5, Scholz and Ziemes 1999, Schou 2000, Schou 2002, Groth and Schou 2002, Grimaud and Rougé 2003). These contributions link new growth theory to the resource economics of the 1970s and the limits-to-growth debate. Since the resource economics of the 1970s is still of central importance, the next section is devoted to a summary before the new literature is taken up.

3. The Wave of Resource Economics in the 1970s

From the literature of the 1970s on non-renewable resources in a macroeconomic framework four contributions published in a symposium issue of *Review of Economic Studies* in 1974 stand out: Dasgupta and Heal (1974), Solow (1974a), and Stiglitz (1974a, 1974b). For the purpose at hand we group these contributions together, notwithstanding they concentrated on partly different aspects and contain far more insight than is visible in this brief account.

3.1. THE DASGUPTA-HEAL-SOLOW-STIGLITZ MODEL

What we may call the Dasgupta-Heal-Solow-Stiglitz model, or D-H-S-S model for short, is a one-sector model with technology and resource constraints described by:

$$Y(t) = F(K(t), L(t), R(t), t), \qquad \partial F/\partial t \geqslant 0, \tag{3}$$

$$\dot{K}(t) = Y(t) - C(t) - \delta K(t), \qquad \delta \geqslant 0, \tag{4}$$

$$\dot{S}(t) = -R(t) \equiv -u(t)S(t), \tag{5}$$
$$L(t) = L_0 e^{nt}, \qquad n \geqslant 0, \tag{6}$$

where $Y(t)$ is aggregate output and $K(t)$, $L(t)$ and $R(t)$ are inputs of capital, labour, and a non-renewable resource (say oil), respectively, at time t. Input of *renewable* natural resources is ignored. The aggregate production function F is neoclassical[17] and has constant returns to scale with respect to K, L, and R. The assumption $\partial F/\partial t \geqslant 0$ represents exogenous technical progress. Further, $C(t)$ is aggregate consumption ($\equiv c(t)L(t)$, where $c(t)$ is per capita consumption), δ denotes a constant rate of capital depreciation (decay),[18] $S(t)$ is the *stock* of the non-renewable resource (e.g. oil reserves), and $u(t)$ is the rate of depletion. Since we must have $S(t) \geqslant 0$ for all t, there is a finite upper bound on cumulative resource extraction:

$$\int_0^\infty R(t)dt \leqslant S(0). \tag{7}$$

Uncertainty and costs of extraction are ignored.[19] There is no distinction between employment $L(t)$ and population. The population growth rate n is assumed constant.

Adding households' preferences and a description of the institutional skeleton (for example competitive markets), the model can be solved. The standard neoclassical (or Solow-Ramsey) growth model (see Barro and Sala-i-Martin 2004) corresponds to the case where neither the production function nor the utility function depends on R or S. This amounts to considering the finiteness of natural resources as economically irrelevant, at least in a growth context. One of the pertinent issues is whether this traditional approach is tenable.

Dasgupta-Heal-Solow-Stiglitz responded to the pessimistic Malthusian views of the Club of Rome (Meadows et al., 1972) by emphasizing that feedback from relative price changes should be taken into account. More specifically they asked the question: what are the conditions needed to avoid a falling level of per capita consumption in the long run in spite of the inevitable decline in resource use? The answer is that there are three ways in which this decline in resource use may be counterbalanced: substitution, resource-augmenting technical progress, and increasing returns to scale. Let us consider each of them in turn (although in practice the three mechanisms tend to be intertwined).

3.2. SUBSTITUTION

By substitution is meant the gradual replacement of the input of the exhaustible natural resource by man-made input, capital. An example might be the substitution of fossil fuel energy by solar, wind, tidal, and wave energy resources; more abundant lower-grade non-renewable resources can be substituted for scarce higher-grade non-renewable resources – and this *will* happen when the scarcity price of these has become sufficiently high; a rise in the price of a mineral may make a synthetic substitute cost-efficient or lead to increased recycling of the mineral; finally, the composition of final output can change towards goods with less material content. The conception is that capital accumulation is at the heart of such processes (though also, the arrival of new technical knowledge may be involved – we come back to this).

Whether capital accumulation can do the job depends critically on the degree of substitutability between K and R. To see this, let the production function F be a Constant-Elasticity-of-Substitution (CES) function with no technical change. That is, suppressing the explicit dating of the variables when not needed for clarity, we have:

$$Y = \left(\alpha K^{\psi} + \beta L^{\psi} + \gamma R^{\psi}\right)^{1/\psi}, \quad \alpha, \beta, \gamma > 0,$$
$$\alpha + \beta + \gamma = 1, \psi < 1, \psi \neq 0. \tag{8}$$

The important parameter is ψ, the substitution parameter. Let p_R denote the cost to the firm per unit of the resource flow and let $\tilde{r}$ be the cost per unit of capital (generally, $\tilde{r} = r + \delta$, where r is the real rate of interest). Then $p_R/\tilde{r}$ is the relative factor price, which may be expected to increase as the resource becomes more scarce. The *elasticity of substitution* between K and R is $[d(K/R)/d(p_R/\tilde{r})]\,(p_R/\tilde{r})/(K/R)$ along an isoquant curve, i.e. the percentage rise in the K-R ratio that a cost-minimizing firm will choose in response to a one per cent rise in the relative factor price, $p_R/\tilde{r}$. For the CES production function this elasticity is a constant $\sigma = 1/(1-\psi) > 0$. Moreover, equation (8) depicts the standard case where the elasticity of substitution between all pairs of production factors is the same.[20]

First, suppose $\sigma > 1$, i.e., $0 < \psi < 1$. Then, for fixed K and L, $Y \to \left(\alpha K^{\psi} + \beta L^{\psi}\right)^{1/\psi} > 0$ when $R \to 0$. In this case of high substitutability the resource is seen to be *inessential* in the sense that it is not necessary for a positive output. That is, from an economic perspective, conservation of the resource is not vital. Instead suppose $\sigma < 1$, i.e., $\psi < 0$. Then output per unit of the resource flow, though increasing when R decreases, is bounded from above. Consequently, the finiteness of the resource inevitably implies doomsday sooner or later (unless, of course, one of the other two salvage mechanisms can prevent it). To see this, keeping K and L fixed, we get

$$\frac{Y}{R} = Y(R^{-\psi})^{1/\psi} = \left[\alpha(\frac{K}{R})^{\psi} + \beta(\frac{L}{R})^{\psi} + \gamma\right]^{1/\psi} \to \gamma^{1/\psi} \text{ for } R \to 0, \tag{9}$$

since $\psi < 0$. In fact, even if K and L are increasing, $\lim_{R\to 0} Y = \lim_{R\to 0}(Y/R)R = \gamma^{1/\psi} \cdot 0 = 0$. Thus, when substitutability is low, the resource is *essential* in the sense that output is nil in its absence.

What about the intermediate case $\sigma = 1$? Although equation (8) is not defined for $\psi = 0$, it can be shown (using L'Hôpital's rule) that $\left(\alpha K^{\psi} + \beta L^{\psi} + \gamma R^{\psi}\right)^{1/\psi} \to K^{\alpha} L^{\beta} R^{\gamma}$ for $\psi \to 0$. This limiting function, a Cobb-Douglas function, has $\sigma = 1$ (corresponding to $\psi = 0$). The interesting aspect of the Cobb-Douglas case is that it is the only case where the resource is essential and at the same time output per unit of the resource is not bounded from above (since $Y/R = K^{\alpha}L^{\beta}R^{\gamma-1} \to \infty$ for $R \to 0$).[21] Under these circumstances it was an open question whether non-decreasing per capita consumption can be sustained. Therefore the Cobb-Douglas case was studied intensively. For example, Solow (1974a) showed the key result that if $n = \delta = 0$, then a necessary and sufficient condition that a constant positive level of consumption can be sustained is that $\alpha > \gamma$. Moreover, this condition seems fairly realistic, since empirically α is several times the size of γ (Nordhaus and Tobin 1972, Neumayer 2000).[22] Solow added the observation that under competitive conditions, the *highest* sustainable level of consumption is obtained when investment in capital exactly equals the resource rent, $R \cdot \partial Y/\partial R$. This result was generalized in Hartwick (1977) and became known as *Hartwick's rule*.

Neumayer (2000) reports that the empirical evidence on the elasticity of substitution between capital and energy is inconclusive. In any case, ecological economists claim the poor substitution case to be much more realistic than the optimistic Cobb-Douglas case, not to speak of the case $\sigma > 1$. This invites considering the role of technical progress.

3.3. TECHNICAL PROGRESS

Solow (1974a) and Stiglitz (1974a,b) analysed the theoretical possibility that resource-saving technological change can overcome the declining resource use that must be expected in the future. In this context the focus is not only on whether a non-decreasing consumption level can be maintained, but also on the possibility of sustained per capita growth in consumption.

New production techniques may raise the efficiency of resource use. For example, Dasgupta (1993) reports that during the period 1900 to the 1960s, the quantity of coal required to generate a kilowatt-hour of electricity fell from nearly seven pounds to less than one pound.[23] Further, technological developments make extraction of lower quality ores cost-effective and make more durable forms of energy economical. Incorporating resource-saving technical progress at the (exogenous) rate $\lambda > 0$, the CES production function reads

$$Y = \left(\alpha K^{\psi} + \beta L^{\psi} + \gamma (A_3 R)^{\psi}\right)^{1/\psi}, \tag{10}$$

where $A_3 = e^{\lambda t}$, assuming, for simplicity, λ to be constant. If the (proportionate) rate of decline of R is kept smaller than λ, then the "effective" resource input is no longer decreasing over time. As a consequence, even if $\sigma < 1$ (the poor substitution case), the finiteness of nature need not be an insurmountable obstacle within any timescale of practical relevance.

Actually, a technology with $\sigma < 1$ *needs* a considerable amount of resource-saving technical progress to obtain compliance with the empirical fact that the income share of natural resources has not been rising (Jones 2002b). When $\sigma < 1$, market forces tend to increase the income share of the factor that is becoming relatively more scarce. Empirically, K/R and Y/R have increased systematically.

However, with a sufficiently increasing A_3, the income share $p_R R/Y$ need not increase in spite of $\sigma < 1$. Similarly, for the model to comply with Kaldor's "stylized facts" (more or less constant growth rates of K/L and Y/L and stationarity of the output–capital ratio, the income share of labour, and the rate of return on capital), we should replace L in equation (10) by $A_2 L$, where A_2 is growing over time. In view of the absence of trend in the rate of return to capital, however, we assume technical progress is on average neither capital-saving nor capital-using, i.e. we do not replace K by $A_1 K$, but leave it as it is.

A concept which has proved extremely useful in the theory of economic growth is the concept of balanced growth. A *balanced growth path* (BGP for short) is defined as a path along which the quantities Y, C, and K change at constant proportionate rates (some or all of which may be negative). It is well known, first, that compliance with Kaldor's "stylized facts" is generally equivalent with existence of a balanced growth path; second, that existence of a balanced growth path requires A_1 to be stationary in the long run, when $\sigma \neq 1$.[24] Of course, one thing is that such a framework may allow for constant growth in per capita consumption – which is more or less what we have seen since the industrial revolution. Another thing is whether such a development will be sustainable for a long time in the future. To come nearer an answer to that question, we need theory about the relation between *endogenous* technical change and non-renewable resources.

Before entering that area, note that the Cobb-Douglas production function is again a convenient intermediate case, in that capital-saving, labour-saving, and resource-saving technical progress are indistinguishable. Hence technical progress can simply be represented by

$$Y = AK^{\alpha}L^{\beta}R^{\gamma}, \tag{11}$$

where "total factor productivity", A, is growing at some constant rate $\tau > 0$. Log-differentiating with respect to time yields the "growth-accounting relation"

$$g_Y = \tau + \alpha g_K + \beta n + \gamma g_R. \tag{12}$$

It is easily shown that along a BGP $g_K = g_Y = g_C \equiv g_c + n$ and, if nothing of the resource is left unutilized forever, $g_R = g_S = -R/S \equiv -u =$ constant, so that equation (12) gives

$$g_c = \frac{1}{1-\alpha}(\tau - \gamma n - \gamma u), \tag{13}$$

since $\alpha + \beta - 1 = \gamma$. Consequently, as observed by Stiglitz (1974a), a positive constant growth rate of c is technologically feasible, if and only if $\tau > \gamma n$. It is also visible from equation (13) that in spite of technical progress being exogenous, there is scope for policy affecting long-run growth to the extent that policy can affect the rate of depletion u in the opposite direction (a property about which we shall have more to say later).

Of course, when speaking of "sustained growth" in K and c, it should not be understood in a narrow physical sense. We have to understand K broadly as "produced means of production" of rising quality and falling material intensity; similarly, c must be seen as a composite of consumer "goods" with declining material intensity over time. This accords with the empirical fact that as income rises, the share of consumption expenditures devoted to agricultural and industrial products declines and the share devoted to services, hobbies, and amusement increases. Although "economic development" is perhaps a more appropriate term, we shall retain standard terminology and speak of "economic growth".

In any event, simple aggregate models like this should be seen as no more than a frame of reference, a tool for thought experiments. At best such models might have some validity as an approximate summary description of a certain period of time. One should be aware that an economy in which the ratio of capital to resource input grows without limit might well enter a phase where technological relations (including the elasticity of substitution) are very different from now.[25]

Dasgupta and Heal (1974) typify a different approach to resource-saving technical change, considering it not as a smooth gradual process, but as something arriving in a discrete once-for-all manner. They envision a future major discovery of, say, how to harness a lasting energy source such that a hitherto essential resource like fossil fuel becomes inessential. The contour of such a "backstop technology" might be currently known, but its practical applicability still awaits a technological breakthrough. The time until the arrival of this breakthrough is uncertain and may well be long. In Dasgupta, Heal, and Majumdar (1977) and Dasgupta, Heal, and Pand (1980) the idea is pursued further, by incorporating costly R&D. The likelihood of the technological breakthrough to appear in a given time interval depends positively on the accumulated R&D as well as the current R&D. It is shown that under certain conditions an index reflecting the probability that the resource becomes unimportant acts like an addition to the utility discount rate and that R&D expenditure begins to decline after some time. This is an interesting example of an early study of *endogenous* technological change. A similar problem has been investigated by Kamien and Schwartz (1978) and Just et al. (2005), using somewhat different approaches.

3.4. INCREASING RETURNS TO SCALE

The third circumstance that might help in overcoming the finiteness of nature is increasing returns to scale. For the CES function with poor substitution ($\sigma < 1$), however, increasing returns to scale, though helping, are not by themselves sufficient to avoid doomsday. To see this, let $Y = \left(\alpha K^{\psi} + \beta L^{\psi} + \gamma R^{\psi}\right)^{\eta/\psi}$, $\eta > 1$. Then

$$\frac{Y}{R^{\eta}} = \left[\alpha\left(\frac{K}{R}\right)^{\psi} + \beta\left(\frac{L}{R}\right)^{\psi} + \gamma\right]^{\eta/\psi} \to \gamma^{\eta/\psi} \text{ for } R \to 0,$$

since $\psi < 0$, when $\sigma < 1$. Hence, even if K and L are increasing, $\lim_{R\to 0} Y = \lim_{R\to 0}(Y/R^{\eta})R^{\eta} = \gamma^{\eta/\psi} \cdot 0 = 0$. In contrast, in the Cobb-Douglas case, equation (11), with $\alpha + \beta + \gamma > 1$, sustained positive per capita growth may be possible. Indeed, as Stiglitz (1974a) noted in a short remark, with increasing returns to scale it is enough that $\tau > (1 - \alpha - \beta)n$, which can be true even if $\tau = 0$.

3.5. SUMMARY OF D-H-S-S

Apart from the just mentioned observation by Stiglitz, the focus of D-H-S-S was on constant returns to scale; and, as in the original Solow-Ramsey growth model, only *exogenous* technical progress was considered. For our purposes we may summarize the D-H-S-S results in the following way. Non-renewable resources do not really matter if the elasticity of substitution between them and man-made inputs is above one. If not, then:

(a) absent technical progress, if $\sigma = 1$, sustainable per capita consumption requires $\alpha > \gamma$ and $n = 0 = \delta$; otherwise, declining per capita consumption is inevitable and this is definitely the prospect, if $\sigma < 1$;
(b) on the other hand, if there is enough resource-saving technical progress, non-decreasing per capita consumption and even growing per capita consumption may be sustained;
(c) population growth (more mouths to feed) exacerbates the drag on growth implied by a declining resource input; indeed, as seen from equation (13), the drag on growth is $\gamma(n + u)/(1 - \alpha)$ along a BGP.

The next sections examine how *endogenizing* technical change may throw new light on the issues, in particular the visions (b) and (c). We shall derive some basic conditions needed for vision (b) to show up. As to point (c), we shall see that the relationship between population growth and economic growth tends to be circumvented when endogenous creation of ideas (generating increasing returns to scale) is considered.

4. Endogenous Growth Theory with Non-Renewable Resources

It is not always recognized that the research of the 1970s on macro implications of essential non-renewable natural resources already laid the groundwork for a theory of endogenous and policy-dependent growth with natural resources. Actually, by extending the D-H-S-S model, Suzuki (1976), Chiarella (1980), Robson (1980), and Takayama (1980) studied how endogenous innovation may affect the prospect of overcoming the finiteness of natural resources. The one-sector model by Suzuki (1976) constitutes an expedient benchmark case.

4.1. AN EXTENDED D-H-S-S MODEL

Suzuki (1976) added *endogenous* technical change to the D-H-S-S model. He insisted that technical innovations are the costly result of intentional R&D. A part of aggregate output is used as R&D investment and results in additional technical knowledge and thereby higher productivity. Aggregate output is

$$Y = A^{\varepsilon} K^{\alpha} L^{\beta} R^{\gamma}, \qquad \varepsilon, \alpha, \beta, \gamma > 0, \;\; \alpha + \beta + \gamma = 1, \tag{14}$$

where A is proportional to the "stock of knowledge". Due to this proportionality we can simply identify A with the stock of knowledge, which increases through R&D investment I_A :

$$\dot{A} = I_A - \delta_A A, \qquad \delta_A \geqslant 0. \tag{15}$$

The interpretation is that the technology for creating new knowledge uses the same inputs as manufacturing, in the same proportions. The parameter δ_A is the (exogenous) rate of depreciation (obsolescence) of knowledge. After consumption and R&D investment, the remainder of output is invested in physical capital:

$$\dot{K} = Y - cL - I_A - \delta_K K, \qquad \delta_K \geqslant 0, \tag{16}$$

where δ_K is the (exogenous) rate of depreciation (decay) of capital. Finally, resource extraction and population growth are described as in equations (5) and (6), respectively. Uncertainty is ignored.

We shall limit our attention to *efficient paths*, i.e. paths such that consumption cannot be increased in some time interval without being decreased in another time interval. Assuming, for simplicity, that $\delta_A = \delta_K = \delta$,[26] the net marginal productivities of A and K are equal if and only if $\varepsilon Y/A - \delta = \alpha Y/K - \delta$, i.e.

$$A/K = \varepsilon/\alpha.$$

Initial stocks, A_0 and K_0, are historically given. Suppose $A_0/K_0 > \varepsilon/\alpha$. Then, initially, the net marginal product of capital is larger than that of knowledge, i.e. capital is relatively scarce. An investing efficient economy will therefore for a while invest only in capital, i.e. there will be a phase where $I_A = 0$. This phase of complete specialization lasts until $A/K = \varepsilon/\alpha$, a state reached in finite time, say at time $\bar{t}$. Hereafter, there is investment in both assets so that their ratio remains equal to the efficient ratio ε/α forever. Similarly, if initially $A_0/K_0 < \varepsilon/\alpha$, then there will be a phase of complete specialization in R&D, and after a finite time interval the efficient ratio $A/K = \varepsilon/\alpha$ is achieved and maintained forever. Thus, for $t > \bar{t}$ it is as if there were only one kind of capital, which we may call "broad capital" and define as $\tilde{K} = K + A = (\alpha + \varepsilon)K/\alpha$. Indeed, substitution of $A = \varepsilon K/\alpha$ and $K = \alpha\tilde{K}/(\varepsilon + \alpha)$ into equation (14) gives

$$Y = \frac{\varepsilon^{\varepsilon}\alpha^{\alpha}}{(\varepsilon + \alpha)^{\varepsilon+\alpha}}\tilde{K}^{\varepsilon+\alpha}L^{\beta}R^{\gamma} \equiv B\tilde{K}^{\tilde{\alpha}}L^{\beta}R^{\gamma}, \qquad \tilde{\alpha} \equiv \alpha + \varepsilon, \tag{17}$$

so that $\tilde{\alpha} + \beta + \gamma > 1$. Further, adding (15) and (16) gives

$$\dot{\tilde{K}} = \dot{A} + \dot{K} = Y - cL - \delta\tilde{K}. \tag{18}$$

Thus, we can proceed with a model based on broad capital, using equations (17), (18), and the usual resource depletion equation (5). Essentially, this model provides a theoretical basis for extending the D-H-S-S model to include increasing returns to scale, thereby offering a simple framework for studying *endogenous* growth with essential non-renewable resources. Groth and Schou (2006) study a similar configuration where the source of increasing returns to scale is not intentional creation of knowledge, but learning as a by-product of investing as in Arrow (1962a) and Romer (1986). Empirically, the evidence furnished by, e.g. Hall (1990) and Caballero and Lyons (1992) suggests that there are quantitatively significant increasing returns to scale with respect to capital and labour or external effects in US and European manufacturing. Similarly, Antweiler and Trefler (2002) examine trade data for goods-producing sectors and find evidence for increasing returns to scale. Whatever the source of increasing returns to scale we shall call a D-H-S-S framework with $\tilde{\alpha} + \beta + \gamma > 1$ an *extended D-H-S-S model.*

Log-differentiating equation (17) with respect to t gives the "growth-accounting equation"

$$g_Y = \tilde{\alpha} g_{\tilde{K}} + \beta n + \gamma g_R. \tag{19}$$

Hence, along a BGP we get, instead of equation (13),

$$(1 - \tilde{\alpha}) g_c + \gamma u = (\tilde{\alpha} + \beta - 1) n. \tag{20}$$

Since $u > 0$, it follows immediately that:

Result (i) A BGP with $g_c > 0$ is technologically feasible only if

$$(\tilde{\alpha} + \beta - 1) n > 0 \quad \text{or} \quad \tilde{\alpha} > 1. \tag{21}$$

This result warrants some remarks from the perspective of new growth theory. In Section 2 we defined *endogenous growth* to be present if sustained positive per capita growth ($g_c > 0$) is driven by some internal mechanism (in contrast to exogenous technology growth). Hence, result (i) tells us that endogenous growth is theoretically possible, if there are either increasing returns to the capital-*cum*-labour input combined with population growth *or* increasing returns to capital (broad capital) itself. At least one of these conditions is required in order for capital accumulation to offset the effects of the inescapable waning of resource use over time. The reasoning of Mankiw (1995) suggests β to be in the neighbourhood of 0.25. And Barro and Sala-i-Martin (2004, p. 110) argue that, given the "broad capital" interpretation of capital, $\tilde{\alpha}$ being around 0.75 accords with the empirical evidence. In view of this, $\tilde{\alpha}$ and β summing to a value above 1 cannot be excluded (but it is, on the other hand, not assured). Hence, $(\tilde{\alpha} + \beta - 1)n > 0$ seems possible when $n > 0$.

We have defined *fully endogenous growth* to be present if the long-run growth rate in per capita output is positive without the support of growth in any exogenous factor. Result (i) shows that only if $\tilde{\alpha} > 1$, is *fully* endogenous growth possible. Although the case $\tilde{\alpha} > 1$ has potentially explosive effects on the economy, if $\tilde{\alpha}$ is not too much above 1, these effects can be held back by the strain on the economy imposed by the declining resource input.[27]

In some sense this is "good news": fully endogenous steady growth is theoretically possible and no knife-edge assumption is needed. As we saw in Section 2, in the conventional framework, without non-renewable resources, fully endogenous growth requires constant returns to the producible input(s) in the growth engine. In our one-sector model the growth engine is the manufacturing sector itself, and without the essential non-renewable resource, fully endogenous growth would require the knife-edge condition $\tilde{\alpha} = 1$ ($\tilde{\alpha}$ being above 1 is excluded in this case, because it would lead to explosive growth in a setting without some countervailing factor). When non-renewable resources are an essential input in the growth engine, they entail a drag on the growth potential. In order to offset this drag, fully endogenous growth requires *increasing* returns to capital.

However, the "bad news" is that even in combination with essential non-renewable resources, an assumption of increasing returns to capital seems too strong and too optimistic. A technology having $\tilde{\alpha}$ just slightly above 1 can sustain *any* per capita growth rate – there is no upper bound on g_c.[28] This appears overly optimistic.

This leaves us with *semi-endogenous* growth as the only plausible form of endogenous growth (as long as n is not endogenous). Indeed, result (i) indicates that semi-endogenous growth corresponds to the case $1 - \beta < \tilde{\alpha} \leqslant 1$. In this case sustained positive per capita growth driven by some internal mechanism is possible, but only if supported by $n > 0$, that is, by growth in an exogenous factor, here population size.

4.2. GROWTH POLICY AND CONSERVATION

Result (i) is about as far as Suzuki's analysis takes us, since his focus is only on whether the technology as such allows the growth rate to be positive or not.[29] That is, he does not study the *size* of the growth rate. A key issue in new growth theory is to explain the size of the growth rate and how it can temporarily or perhaps permanently be affected by economic policy. The simple growth-accounting relation equation (20) immediately shows:

Result (ii) Along a BGP, policies that decrease (increase) the depletion rate u (and only such policies) will increase (decrease) the per capita growth rate (here we presuppose $\tilde{a} < 1$, the plausible case).

This observation is of particular interest in view of the fact that changing the perspective from exogenous to endogenous technical progress implies bringing a source of numerous market failures to light. On the face of it, the result seems to run against common sense. Does high growth not imply *fast* depletion (high u)? Indeed, the answer is affirmative, but with the addition that exactly because of the fast depletion such high growth will only be temporary – it carries the seeds to its own obliteration. For faster sustained growth there must be sustained slower depletion. The reason for this is that with protracted depletion, the rate of decline in resource input becomes smaller; hence, so does the drag on growth caused by this decline.

As a statement about policy and long-run growth, result (ii) is a surprisingly succinct conclusion. It can be clarified in the following way. For policy to affect long-run growth, it must affect a linear differential equation linked to the basic goods sector in the model (Romer 1995). In the present framework the resource depletion relation,

$$\dot{S} = -uS,$$

is such an equation. In balanced growth $g_S = -R/S \equiv -u$ is constant so that the proportionate rate of decline in R must comply with, indeed be equal to, that of S. Through the growth accounting relation equation (19), given u, this fixes g_Y and $g_{\tilde{K}}$ (equal in balanced growth), hence also $g_c = g_Y - n$. The conventional wisdom in the endogenous growth literature is that interest income taxes impede economic growth and investment subsidies promote economic growth. Interestingly, this is not so when non-renewable resources are an essential input in the growth engine (which is here the manufacturing sector itself). Then, generally, only those policies that interfere with the depletion rate u in the long run (like a profits tax on resource-extracting companies or a time-dependent tax on resource use) can affect long-run growth. This is further explored in Groth and Schou (2006). It is noteworthy that this long-run policy result holds whether $g_c > 0$ or not and whether growth is exogenous, semi-endogenous, or fully endogenous.[30] The general conclusion is that with non-renewable resources entering the growth-generating sector in an essential way, conventional policy tools receive a different role and there is a role for new tools (affecting long-run growth through affecting the depletion rate).

4.3. FURTHER IMPLICATIONS

In order to be more specific we introduce household preferences and a "social planner". The resulting resource allocation will coincide with that of a decentralized economy with appropriate subsidies and taxes. As in Stiglitz (1974a), let the utilitarian social planner optimize

$$U_0 = \int_0^\infty \frac{c(t)^{1-\theta} - 1}{1-\theta} L(t) e^{-\rho t} dt, \qquad \theta > 0, \rho \geqslant n \geqslant 0, \tag{22}$$

subject to the constraints given by technology (equations (17), (18), and (5)) and initial conditions. Here, θ is the (numerical) elasticity of marginal utility (desire for consumption smoothing) and ρ is a constant rate of time preference (impatience).[31]

Using the Pontryagin Maximum Principle, the first order conditions for this problem lead to, first, the *Ramsey rule*,[32]

$$g_c = \frac{1}{\theta}\left(\frac{\partial Y}{\partial \tilde{K}} - \delta - \rho\right) = \frac{1}{\theta}\left(\tilde{\alpha}\frac{Y}{\tilde{K}} - \delta - \rho\right), \tag{23}$$

second, the *Hotelling rule*,[33]

$$\frac{d(\partial Y/\partial R)}{dt} = \frac{\partial Y}{\partial R}\left(\frac{\partial Y}{\partial \tilde{K}} - \delta\right) = \gamma \frac{Y}{R}\left(\tilde{\alpha}\frac{Y}{\tilde{K}} - \delta\right). \tag{24}$$

The first rule says: as long as the net return on investment in capital is higher than the rate of time preference, one should let current c be low enough to allow positive net saving (investment) and thereby higher consumption in the future. The second rule is a no-arbitrage condition saying that the return ("capital gain") on leaving the marginal unit of the resource in the ground must equal the return on extracting and using it in production and then investing the proceeds in the alternative asset (reproducible capital).[34]

Using the Cobb-Douglas specification, we may rewrite the Hotelling rule as $g_Y - g_R = \tilde{\alpha}Y/\tilde{K} - \delta$. Along a BGP $g_Y = g_C = g_c + n$ and $g_R = -u$, so that the Hotelling rule combined with the Ramsey rule gives

$$(\theta - 1)g_c - u = n - \rho. \tag{25}$$

This linear equation in g_c and u combined with the growth-accounting relationship equation (20) constitutes a linear two-equation system in the growth rate and the depletion rate. The determinant of this system is $D \equiv 1 - \tilde{\alpha} - \gamma + \theta\gamma$. We assume $D > 0$, which seems realistic and is in any case necessary (and sufficient) for stability.[35] Then

$$g_c = \frac{(\tilde{\alpha} + \beta + \gamma - 1)n - \gamma\rho}{D}, \quad \text{and} \tag{26}$$

$$u = \frac{[(\tilde{\alpha} + \beta - 1)\theta - \beta]n + (1 - \tilde{\alpha})\rho}{D}. \tag{27}$$

Interesting implications are:

Result (iii) If there is impatience ($\rho > 0$), then even when a non-negative g_c is technologically feasible equation (21) is satisfied, a negative g_c can be optimal and stable.

Result (iv) Population growth is *good* for economic growth. In its absence, when $\rho > 0$, we get $g_c < 0$ along an optimal BGP; if $\rho = 0$, $g_c = 0$ when $n = 0$.

Result (v) There is never a scale effect on the growth rate.

Result (iii) reflects that utility discounting and consumption smoothing weaken the "growth incentive". Result (iv) is completely contrary to the conventional (Malthusian) view and the learning from the D-H-S-S model. The point is that two offsetting forces are in play. On the one hand, higher n means more mouths to feed and thus implies a drag on per capita growth (Malthus). On the other hand, a growing labour force is exactly what is needed in order to exploit the benefits of increasing returns to scale (anti-Malthus).[36] And in the present framework this dominates the first effect.[37] This feature might seem to be contradicted by the empirical finding that there is no robust correlation between g_c and population growth in cross-country regressions (Barro and Sala-i-Martin 2004, Chap. 12). However, the proper unit of observation in this context is not the individual country. Indeed, as argued in Section 2.2, in an internationalized world with technology diffusion a positive association between n and g_c as in equation (26) should not be seen as a prediction about individual countries, but rather as pertaining to larger regions, perhaps the global economy. In any event, the second part of result (iv) is a dismal part – in view of the projected long-run stationarity of world population (United Nations 2005).

A somewhat surprising result appears if we imagine (unrealistically) that $\tilde{\alpha}$ is sufficiently above one to make D a negative number. If population growth is absent, $D < 0$ is in fact needed for $g_c > 0$ along a BGP. However, $D < 0$ implies instability. Hence this would be a case of an instable BGP with fully endogenous growth.[38]

As to result (v), it is noteworthy that the absence of a scale effect on growth holds for *any* value of $\tilde{\alpha}$, including $\tilde{\alpha} = 1$.[39]

A pertinent question now is: are the above results just an artifact of the one-sector set-up? This leads us to consider two-sector models.

5. Models with a Separate R&D Sector

5.1. THE STANDARD APPROACH

The results (i), (ii), (iii), and (v) above (and partly also (iv)) differ from most of the new growth literature,[40] including most of the contributions that deal explicitly with non-renewable resources and endogenous growth (Jones and Manuelli 1997; Aghion and Howitt 1998 (Chap. 5); Scholz and Ziemes 1999; Schou 2000; Schou 2002; Grimaud and Rougé 2003). These contributions extend the first-generation two-sector endogenous growth models referred to in Section 2, by including a non-renewable resource as an essential input in the manufacturing sector. The non-renewable resource does not, however, enter the R&D or educational sector in these models (not even indirectly in the sense of physical capital produced in the manufacturing sector being used in the R&D sector). As we shall now see, this is the reason that these models give results quite similar to those from conventional endogenous models without non-renewable resources.

The following two-sector framework is a prototype of the afore-mentioned contributions:

$$Y = A^{\varepsilon} K^{\alpha} L_Y^{\beta} R^{\gamma}, \quad \varepsilon, \alpha, \beta, \gamma > 0, \ \alpha + \beta + \gamma = 1, \tag{28}$$

$$\dot{K} = Y - cL - \delta K, \quad \delta \geqslant 0,$$

$$\dot{A} = \bar{\mu} L_A, \quad \bar{\mu} = \mu A, \quad \mu > 0, \tag{29}$$

$$\dot{S} = -R,$$

$$L_Y + L_A = L, \text{ constant.}$$

Unlike in the previous model, additions to society's "stock of knowledge", A, are now produced in a separate sector, the R&D sector, with a technology different from that in manufacturing. The only input in the R&D sector is labour (thus taking to the extreme the feature that this sector is likely to be relatively intensive in human capital). The individual research lab, which is "small" in relation to the economy as a whole, takes R&D productivity, $\bar{\mu}$, as given. At the economy-wide level, however, this productivity depends positively on the stock of technical knowledge in society, A (this externality is one of several reasons that the existence of endogenous technical change implies market failures). Usually, there is no depreciation of knowledge, i.e. $\delta_A = 0$. Aggregate employment in the R&D sector is L_A. Total employment, L, in the economy is the sum of L_A and employment, L_Y, in the manufacturing sector. In that sector, the firms take A as given and the technology they face at the micro level may involve different capital-good varieties and qualities. There are many interesting details and disparities between the models concerning these aspects as well as the specifics of the market structure and the policy questions considered. Yet, whether we think of the "increasing variety" models (or Romer-style models to which Scholz and Ziemes 1999 and Schou 2002 belong) or the "increasing quality models" (or quality ladder models to which Aghion and Howitt 1998 and Grimaud and Rougé 2003 belong), at the aggregate level these models end up with a formal structure basically like that above.[41] The accumulation-based growth models by Jones and Manuelli (1997) and Schou (2000) are in one respect different – we shall return to this.

Two key features emphasized by new growth theory are immediately apparent. First, because technological ideas – sets of instructions – are *non-rival*, what enters both in the production function for Y and that for $\dot{A}$ is *total* A. This is in contrast to the *rival* goods: capital, labour, and the resource flow. For example, a given unit of labour can be used no more than one place at a time. Hence, only a fraction of the labour force enters manufacturing, the remaining fraction entering R&D. Second, there is a tendency for *increasing returns to scale* to arise when knowledge is included in the total set of inputs. At least when we ignore externalities, the well-known replication argument gives reason to expect constant returns to scale with respect to the *rival* inputs (here K, L_Y, and R in the manufacturing sector and L_A in R&D). Consequently, as we double these rival inputs and *also* double the amount of knowledge, we should expect more than a doubling of Y and $\dot{A}$. An additional key feature of new growth theory, apparent when the above technology description is combined with assumptions about preferences and market structure, is the emphasis on incentives as driving R&D investment. When the resource becomes more scarce and its price rises, the value of resource-saving knowledge increases and R&D is stimulated.[42]

Using the principle of growth accounting on equation (28), taking $n = 0$ into account, we get, along a BGP,[43]

$$(1-\alpha)g_c = \varepsilon g_A - \gamma u, \tag{30}$$

where

$$g_A = \mu \ell_A L, \qquad \ell_A \equiv \frac{L_A}{L}, \text{ constant.}$$

We have $g_A > 0$ if $\ell_A > 0$. The essential non-renewable resource implies a drag on the growth of consumption. Yet, by sufficient conservation of the resource (implying a small $u \equiv R/S$) it is always possible to obtain $g_c > 0$. And it is possible to increase g_c without decreasing u, simply by increasing ℓ_A. These two last conclusions have a quite different flavour compared to the results (i) and (ii) from the extended D-H-S-S model.

The fraction, ℓ_A, of the labour force in R&D will depend on parameters such as α, ε, μ, and those describing preferences and the allocation device, whether this is the market mechanism in a decentralized economy or the social planner in a centralized economy. To be specific, let us again consider a social planner and the criterion function (22). Along a BGP we get once more equation (25) (from the Ramsey rule and the Hotelling rule). Further, efficient allocation of labour across the two sectors and across time leads to $\ell_A = 1 - \beta u/(\varepsilon \mu L)$. Combining this with equations (30) and (25) we find, along a BGP,

$$\begin{aligned} \ell_A &= \frac{\varepsilon \mu L(\beta + \theta \gamma) - \beta(1-\alpha)\rho}{\varepsilon \mu L \theta(1-\alpha)}, \\ g_c &= \frac{\varepsilon \mu L - (1-\alpha)\rho}{\theta(1-\alpha)}, \quad \text{and} \\ u &= \frac{(\theta - 1)\varepsilon \mu L + (1-\alpha)\rho}{\theta(1-\alpha)}. \end{aligned}$$

This is an example of fully endogenous growth: given $(1-\theta)\varepsilon\mu L < (1-\alpha)\rho < \varepsilon\mu L$,[44] per capita growth is positive along a BGP without support of growth in any exogenous factor. A caveat is that this result relies on the knife-edge assumption that the growth engine (the R&D sector) has exactly constant returns to the producible input(s), here A. The problematic scale effect on growth ($\partial g_c/\partial L > 0$) crops up again (although often hidden by the labour force being normalized to one). Indeed, this is why these models assume a constant labour force; with $n > 0$ the growth rate will be forever rising. In any event, contrary to the implication of equation (26), sustained positive growth is conceivable without population growth and whether $\rho = 0$ or $\rho > 0$.

Overall, we have a more optimistic perspective than in the extended D-H-S-S model. Indeed, the conclusions are quite different from the results (i), (ii), and (v) above (and partly also different from (iv)). The conclusions are, however, pretty much in conformity with those of the fully endogenous growth models without non-renewable resources. With the exception of the scale effect on growth we get similar results in the model by Jones and Manuelli (1997). They consider an economy with a sector producing consumption goods with labour, capital, and the non-renewable resource and a sector producing capital goods with only capital (not even labour). The model by Schou (2000) is a Lucas-style human-capital-based model extended with a non-renewable resource entering only the manufacturing sector (with the addition of pollution from this resource). Since in both models it is the accumulation of a *rival* good that drives growth, the scale effect on growth does not appear, but this is the only difference in relation to the questions considered here.

The explanation of the optimistic results in all these models is that the growth-generating sector is presumed not to depend on the non-renewable resource (neither directly nor indirectly). In reality, however, most sectors, including educational institutions and research laboratories, use fossil fuels for heating and transportation purposes, or at least they use indirectly minerals and oil products via the machinery, computers etc. they employ. The extended D-H-S-S model in the previous section *did* take this dependency of the growth engine (in that model the manufacturing sector itself) on the natural resource into account and therefore gave substantially different results. In the next section we shall see that a two-sector model with the resource entering also the R&D sector leads to results similar to those of the extended D-H-S-S model from Section 4, but quite different from those of the above two-sector model.

5.2. GROWTH-ESSENTIAL NON-RENEWABLE RESOURCES

When a natural resource is an essential input (directly or indirectly) in the growth-engine, we shall call the resource *growth-essential.*

5.2.1. *The resource as input in both sectors*

Extending the above two-sector framework as in Groth (2005), we consider the set-up:

$$
\begin{aligned}
Y &= A^{\varepsilon} K^{\alpha} L_Y^{\beta} R_Y^{\gamma}, \quad \varepsilon, \alpha, \beta, \gamma > 0, \;\; \alpha + \beta + \gamma = 1, && (31)\\
\dot{K} &= Y - cL - \delta K, \quad \delta \geqslant 0, && (32)\\
\dot{A} &= \bar{\mu} L_A^{\eta} R_A^{1-\eta}, \quad \bar{\mu} = \mu A^{\varphi}, \quad \mu > 0, \;\; 0 < \eta < 1, && (33)\\
\dot{S} &= -R, && (34)\\
L_Y + L_A &= L = L(0)e^{nt}, \quad n \geqslant 0, && (35)\\
R_Y + R_A &= R. && (36)
\end{aligned}
$$

There are three new features. First, only a fraction of the resource flow R is used in manufacturing, the remainder being used as an essential input in R&D activity. Second, the knowledge elasticity, φ, of research productivity is allowed to differ from one; as argued in the section on the Jones critique, even $\varphi < 0$ should not be excluded *a priori*. Third, population growth is not excluded.

Along a BGP, using the principle of growth accounting on equation (31) yields

$$(1-\alpha)g_c = \varepsilon g_A - \gamma(n+u). \tag{37}$$

Applying the same principle on the R&D equation (33) (after dividing by A and presupposing the R&D sector is active) and assuming balanced growth we get, after substituting into equation (37),

$$(1-\alpha)g_c = \left(\frac{\varepsilon\eta}{1-\varphi} - \gamma\right)n - \left(\frac{\varepsilon(1-\eta)}{1-\varphi} + \gamma\right)u. \tag{38}$$

Since $u > 0$, from this[45] follows that a BGP with $g_c > 0$ is technologically feasible only if

$$\varphi < 1 + \frac{\varepsilon(1-\eta)}{\gamma} \text{ and either } (n > 0 \text{ and } \varepsilon\eta > (1-\varphi)\gamma) \text{ or } \varphi > 1.$$

Naturally, the least upper bound for φ's that allow non-explosive growth is here higher than when the resource is not a necessary input in the R&D sector. We also see that for the technology to allow steady positive per capita growth, *either* φ must be above one *or* there must be population growth (to exploit increasing returns to scale) *and* an elasticity of Y with respect to knowledge large enough to overcome the drag on growth caused by the inevitable decline in resource use. Not surprisingly, in the absence of population growth, sustained per capita growth requires a higher elasticity of research productivity with respect to knowledge than when the growth engine does not need the resource as an input. The "standard" two-sector model of the previous section relied on the aggregate invention production function having exactly constant returns (at least asymptotically) to produced inputs, that is, $\varphi = 1$. Slightly increasing returns with respect to A would in that model lead to explosive growth, whereas slightly decreasing returns lead to growth petering out. Interestingly, when the resource is growth-essential, the case $\varphi = 1$ loses much of its distinctiveness. Yet, the "bad news" for fully endogenous growth is again that $\varphi > 1$ seems to be a too optimistic and strong assumption. The reason is similar to that given in Section 4.1 for doubting that $\tilde{\alpha} > 1$, namely that whenever a given technology has $\varphi > 1$, it can sustain *any* per capita growth rate no matter how high – a rather suspect implication. Thus, once more we are left with semi-endogenous growth ($\varphi \leqslant 1$) as the only appealing form of endogenous growth (as long as n is exogenous).

In parallel to result (ii) above, equation (38) shows that when $\varphi < 1$, only policies that decrease the depletion rate u along a BGP, can increase the per capita growth rate g_c. For example, embedding the just described technology in a Romer (1990)-style market structure, Groth (2006) shows that a research subsidy, an interest income tax, and an investment subsidy do not affect long-run growth whereas taxes that impinge on resource extraction do. The point is that whatever market forms might embed the described technology and whatever policy instruments are considered, the growth-accounting relation (38) *must* hold (given the assumed Cobb-Douglas technologies).

Let us again consider a social planner and the criterion function (22). Then, along a BGP we have once more equation (25) (from the Ramsey rule and the Hotelling rule). Combining this with equation (38) we find, along a BGP,

$$g_c = \frac{\varepsilon n - [\varepsilon(1-\eta) + (1-\varphi)\gamma]\,\rho}{\tilde{D}}, \quad \text{and}$$
$$u = \frac{\left[(\theta-1)\varepsilon - \tilde{D}\right] n + (1-\varphi)(1-\alpha)\rho}{\tilde{D}},$$

where $\tilde{D} \equiv (1-\varphi)(\beta+\theta\gamma) + (\theta-1)\varepsilon(1-\eta)$ is assumed positive (this seems to be the empirically relevant case and it is in any event necessary, though not sufficient, for stability).[46] We see that in the plausible case $\varphi < 1 + \varepsilon(1-\eta)\gamma$ the analogy of the results (iii), (iv), and (v) from the extended D-H-S-S model of Section 4 goes through.[47]

The conclusion is that when a non-renewable resource is an essential input in the R&D sector, quite different and more pessimistic conclusions arise compared to those of the previous section. Sustained growth without increasing research effort (i.e. without $n > 0$) now requires $\varphi > 1$ in contrast to $\varphi = 1$ in the previous section. In the realistic case $\varphi < 1$, policies aimed at stimulating long-run growth have to go via resource conservation.

5.2.2. *Capital in the R&D sector*

The results are essentially the same in the case where the resource is a direct input only in manufacturing, but the R&D sector uses capital goods (apparatus and instruments) produced in the manufacturing sector. Thus, indirectly the resource is an input also in the R&D sector, hence still growth-essential. The model is:

$$
\begin{aligned}
Y &= A^{\varepsilon} K_Y^{\alpha} L_Y^{\beta} R^{\gamma}, \quad \varepsilon, \alpha, \beta, \gamma > 0, \;\; \alpha+\beta+\gamma = 1, && (39)\\
\dot{K} &= Y - cL - \delta K, \quad \delta \geqslant 0, && \\
\dot{A} &= \bar{\mu} K_A^{1-\eta} L_A^{\eta}, \quad \bar{\mu} = \mu A^{\varphi}, \quad \mu > 0, \quad 0 < \eta < 1, && (40)\\
\dot{S} &= -R, && \\
K_Y + K_A &= K, && (41)\\
L_Y + L_A &= L = L(0)e^{nt}, \quad n \geqslant 0. &&
\end{aligned}
$$

Possibly, $1 - \eta < \alpha$ (since the R&D sector is likely to be relatively intensive in human capital), but for our purposes here this is not crucial.

Using the growth accounting principle on equation (39) again gives equation (37) along a BGP. Applying the same principle on the R&D equation (40) (presupposing the R&D sector is active) and assuming balanced growth, we find

$$(1-\varphi)g_A = (1-\eta)g_K + \eta n = (1-\eta)g_c + n, \tag{42}$$

in view of $g_K = g_C = g_c + n$. This shows that existence of a BGP with positive growth requires $\varphi < 1$.[48] Both K and A are essential producible inputs in the two sectors; hence, the two sectors together make up the growth engine.

Substituting equation (42) into equation (37) yields

$$[(1-\varphi)(1-\alpha)-\varepsilon(1-\eta)]\, g_c = [\varepsilon-(1-\varphi)\gamma]\, n-(1-\varphi)\gamma u. \qquad (43)$$

Since $u>0$, we see that a BGP with $g_c>0$ is technologically feasible only if, in addition to the requirement $\varphi<1$,

$$\text{either } (\varepsilon>(1-\varphi)\gamma \text{ and } n>0) \text{ or } \varepsilon>\frac{(1-\varphi)(1-\alpha)}{1-\eta}.$$

That is, given $\varphi<1$, the knowledge elasticity of manufacturing output should be high enough. These observations generalize result (i) from the extended D-H-S-S model and also result (ii), when we (plausibly) assume $\varepsilon<(1-\varphi)(1-\alpha)/(1-\eta)$, which corresponds to $\tilde{\alpha}<1$ in the one-sector model. The combined accumulation of K and A drives growth, possibly with the help of population growth.

Again, let us consider a social planner and the criterion function (22). Along a BGP we get once more equation (25) (from the Ramsey rule and the Hotelling rule). Combining this with equation (43) yields, along a BGP,

$$g_c = \frac{\varepsilon n-(1-\varphi)\gamma\rho}{D^*}, \qquad \text{and}$$

$$u = \frac{[(\theta-1)\varepsilon-D^*]\, n+[(1-\varphi)(1-\alpha)-\varepsilon(1-\eta)]\,\rho}{D^*},$$

where $D^*\equiv(1-\varphi)(\beta+\theta\gamma)-\varepsilon(1-\eta)$ is assumed positive. The results (iii), (iv), and (v) from the extended D-H-S-S model immediately go through.

Thus, also when the non-renewable resource is only indirectly growth-essential, do we get conclusions in conformity with those in the previous subsection, but quite different from those of standard endogenous growth models with non-renewable resources entering only the manufacturing sector. This is somewhat at variance with the reflections on growth and non-renewable resources in Aghion and Howitt (1998). They compare their two-sector Schumpeterian approach (which in this context is equivalent to what was above called "the standard approach") with a one-sector AK model extended with an essential non-renewable resource and no population growth (which is equivalent to the extended D-H-S-S model with $\alpha=1$ and $n=0$). Having established that sustained growth is possible in the first approach, but not in the second, they ascribe this difference to "the ability of the Schumpeterian approach to take into account that the accumulation of intellectual capital is 'greener' (in this case, less resource intensive) than the accumulation of tangible capital" (p. 162). However, as the above example shows, even allowing the R&D sector to be "greener" than the manufacturing sector, we may easily end up with AK-style results. The crucial distinction is between models where the non-renewable resource is growth-essential – directly or indirectly – and models where it is not. To put it differently: by not letting the resource enter the growth engine (not even indirectly), Aghion and Howitt's "Schumpeterian approach" seems biased toward sustainability.

5.2.3. *The case of limited substitutability in the R&D sector*

One might argue that, at least in the R&D sector, the elasticity of substitution between labour (research) and other inputs must be low. Hence, let us consider the limiting case of zero substitutability in the models of the two previous subsections. First, we replace equation (33) in the model of Section 5.2.1 by

$$\dot{A} = \mu A^{\varphi} \min\left(L_A, A^{\psi} R_A\right), \qquad \psi > 0.$$

Then, along any efficient path with $g_A > 0$ we have $L_A = A^{\psi} R_A$ so that $g_A = \mu A^{\varphi-1} L_A = \mu A^{\varphi+\psi-1} R_A$. Log-differentiating this with respect to t and setting $\dot{g}_A = 0$ gives, along a BGP, $(\varphi - 1)g_A + n = 0 = (\varphi + \psi - 1)g_A - u$. Since $n \geqslant 0$ and $u > 0$, $1 - \psi < \varphi \leqslant 1$ is required (if $\varphi > 1$, growth becomes explosive). In the generic case $\varphi < 1$, $g_A = n/(1 - \varphi)$ so that $g_A > 0$ requires $n > 0$; we end up with

$$\begin{aligned} g_c &= \frac{\varepsilon - \gamma\psi}{(1-\alpha)(1-\varphi)} n, \\ u &= \frac{\varphi + \psi - 1}{1 - \varphi} n. \end{aligned}$$

Thus, both the per capita consumption growth rate and the depletion rate u along a BGP are in this case technologically determined. As an implication, preferences and economic policy can have only level effects, not long-run growth effects. If $n = 0$, no BGP with $g_c > 0$ exists in this case.

The singular case $\varphi = 1$ is different. This is the only case where there is scope for preferences and policy to affect long-run growth. Indeed, in this case, where $n = 0$ is needed to avoid a forever increasing growth rate, along a BGP we get $g_c = (\varepsilon - \gamma\psi)\mu L_A$ and $u = \psi\mu L_A$.

We get similar results if in the model of Section 5.2.2 we replace equation (40) by

$$\dot{A} = \mu A^{\varphi} \min\left(K_A, A^{\psi} L_A\right), \qquad \psi > 0.$$

Along any efficient path with $g_A > 0$, now $K_A = A^{\psi} L_A$ so that $g_A = \mu A^{\varphi-1} K_A = \mu A^{\varphi+\psi-1} L_A$. Log-differentiating this with respect to t and setting $\dot{g}_A = 0$ gives, along a BGP, $(\varphi - 1)g_A + g_K = 0 = (\varphi + \psi - 1)g_A + n$. Since $n \geqslant 0$, $\varphi \leqslant 1 - \psi$ is required (if $\varphi > 1 - \psi$, growth becomes explosive). In the generic case $\varphi < 1 - \psi$, both the depletion rate u and the per capita consumption growth rate become technologically determined:

$$\begin{aligned} g_c &= \frac{\psi}{1 - \varphi - \psi} n, \\ u &= \frac{\varepsilon - \beta\psi - \gamma(1-\varphi)}{(1 - \varphi - \psi)\gamma} n, \end{aligned}$$

where the inequalities $n > 0$ and $\varepsilon > \beta\psi + \gamma(1 - \varphi)$ are presupposed. If $n = 0$, no BGP with $g_A > 0$ exists in this case.

Only in the singular case $\varphi = 1 - \psi$ can preferences and policy affect long-run growth. Indeed, in this case, where $n = 0$ is needed to avoid a forever increasing growth rate, along a BGP we find $g_c = \psi\mu L_A$ and $u = (\varepsilon - (1-\alpha)\psi)\mu L_A$, where $\varepsilon > (1-\alpha)\psi$ is presupposed.

To conclude, with zero substitution between the production factors in the R&D sector, one "degree of freedom" is lost. As an implication, in the generic case there is no scope for preferences and policy affecting growth. Only in a knife-edge case can preferences and policy affect growth. Thus, the robust case is in this regard in conformity with semi-endogenous growth models without non-renewable resources à la Jones (1995b), and the non-robust case is in conformity with fully endogenous growth models without non-renewable resources à la Romer (1990).

6. Discussion

New growth theory suggests that costly innovation is the key factor in overcoming the inevitable decline in use of non-renewable resources. Yet how much innovations, together with accumulation of capital, can achieve, depends on the returns to producible inputs, including technical knowledge. We have argued for the neoclassical view that diminishing returns are the most likely case. Then the growing technical knowledge that is needed for continued economic growth requires sustained growth in research effort to countervail the diminishing returns. With a rising population there is scope for a rising number of researchers and the growth prospects seem relatively fine. However, the general conception is that economic and cultural conditions are likely to put an end to population growth within 40–80 years and as early as 20–25 years in the now more developed regions (United Nations 2005). Thus, according to the theory above we should expect a slowdown of long-run per capita growth.

There *are* counteracting forces though. The UN prediction that growth in world population will come to a halt does not necessarily mean that the n relevant for the technological frontier will be approaching zero equally soon. Even a stationary population does not preclude rising research intensity and educational attainment for a quite long time (Jones 2002a). Longevity is apt to help and so are improved institutional structures. Further, as Solow (1994) remarked "there is probably an irreducibly exogenous element in the research and development process, at least exogenous to the economy. [...] the 'production' of new technology may not be a simple matter of inputs and outputs" in the way our models have assumed.

Overall, the abstract character and the insufficient empirical underpinnings of the models call for caution with regard to the big question of limits to growth. But at least it seems safe to infer that endogenizing technical change substantiates the old view that if non-renewable resources are essential, they will ultimately cause a drag on growth. That is, growth ends up smaller than otherwise. In this context one should remember that even if exponential growth ceases, this need not imply absence of growth altogether. Leaving the confines of balanced growth opens up for considering a whole range of less-than-exponential, yet regular, growth paths (with complete stagnation as the limiting case).[49]

There are several complicating factors the above analysis has left aside; and many issues at the interface of resource economics and new growth theory have not been considered. Here we list some of these.

1. *Extraction costs and an enriched Hotelling time pattern of energy prices.* Our analysis of endogenous technical change with non-renewable resources share two empirically questionable features with the D-H-S-S model and the original Hotelling (1931) principle. These are the predictions that real resource prices should have a positive trend and resource consumption should have a negative trend. The empirical evidence stretching over more than a century does not confirm this (Nordhaus 1992, Smil 1994, Krautkraemer 1998, and Jones 2002b). Tahvonen and Salo (2001) therefore propose a different approach where there is a gradual transition from (non-essential) non-renewable energy forms to renewable energy forms (hydropower, wind-energy, solar energy, biomass and geothermal energy). There are extraction costs associated with non-renewable energy sources and these costs are decreasing in remaining reserves and extraction knowledge (a by-product of cumulative extraction experience). Know-how relevant to renewable energy sources is formed as a by-product of physical capital investment. This makes renewable energy forms more and more cost-efficient and an asymptotic AK structure in line with Rebelo (1991) arises, thus making sustained growth feasible. A possible endogenous outcome of all this is a long period of declining resource prices and rising use of non-renewables followed by a shorter period with Hotelling-style trends before finally the renewable resources completely take over.

2. *CES technology with $\sigma < 1$. Induced bias.* We have concentrated on one- and two-sector models with Cobb-Douglas technology. In this setting the elasticity of factor substitution, σ, is 1 and technological progress is automatically resource-augmenting. Perhaps this may not be as serious a restriction as one might think at first. Jones (2005) provides microfoundations for the production function being Cobb-Douglas in the long run, though the short-term elasticity of substitution is likely to be less than one. Yet, it is worth considering the possibility that $\sigma < 1$ also in the long term. In that case technical progress must in the long run be resource-augmenting and labour-augmenting, but not capital-augmenting, to allow for a BGP at least roughly consistent with the empirical evidence. Building on Acemoglu (2003), Di Maria and Valente (2006) show how such bias in technical progress may come about endogenously in a model where both the rate and the direction of technical change are governed by profit incentives. In a similar vein, André and Smulders (2004), extending Smulders and Nooij (2003), demonstrate how induced bias may lead to an U-shaped time pattern for energy prices relative to wages and an inverted U-shaped pattern for energy use.

Bretschger and Smulders (2006) consider an R&D-based growth model with *two* manufacturing sectors, a "traditional" sector and a "high-tech" sector, both with CES production functions where the elasticity of substitution between intermediate (non-durable) goods and the non-renewable resource is less than one.

Provided the elasticity of substitution in the high-tech sector is the highest (and some further conditions), relative price changes shifts consumption demand gradually towards the high-tech sector, and this helps overcoming the decline in the resource input. Yet, what makes sustained *growth* possible is the presumed unitary elasticity of substitution between a man-made input, in this case *knowledge*, and the resource. Thus, the general principle from Section 3 survives.

3. *Amenity value.* In addition to being valued as inputs in production, natural resources may be assets of value in their own right (amenity value, an argument in the utility function). Although this concern seems more prevailing in relation to environmental goods of a *renewable* resource character, Krautkraemer (1985) and Heal (1998) also study its implications in the context of non-renewable resources and its relation to sustainable development.

4. *Polluting non-renewable resources.* There often are negative externalities associated with the use of non-renewable resources, global warming being a glaring example. In the Suzuki (1976) paper there is a companion model to the one considered in Section 4.1. That companion model links the greenhouse problem to the non-renewable nature of fossil fuels. This is further developed in Sinclair (1994) and Groth and Schou (2006). An analysis closer to the global carbon cycle models of the climatologists is contained in Farzin and Tahvonen (1996). Schou (2000) and Schou (2002) study other aspects of (flow) pollution from use of non-renewable resources.

5. *Other issues.* We have completely passed over the role of *uncertainty* as to size of reserves, outcome of R&D activity, future technology, prices and interest rates. The reader is referred to, e.g. Chichilnisky et al. (1998), Weitzman (1998b, 2001), and Just et al. (2005). The problem of the *non-existence of a complete set of forward markets* (and therefore markets for contingent sales) and the associated stability problems were already intensively discussed in Dasgupta and Heal (1979). The *empirics* of resource scarcity are surveyed in Krautkraemer (1998).

7. Summary and Conclusion

To the extent that non-renewable resources are necessary inputs in production, sustained growth requires the presence of resource-augmenting technical progress. New growth theory has deepened our understanding of mechanisms that influence the amount and direction of technical change. Applying new growth theory to the field of resource economics and the problems of sustainability yields many insights. The findings emphasized in this article are the following. (1) As expected, in view of the inevitable decline in resource input, whether technical change is exogenous or endogenous, essential non-renewable resources ultimately imply a drag on growth. (2) By calling attention to the non-rivalrousness of technical knowledge, new growth theory has circumvented the relationship between population growth and economic growth; contrary to the teaching implied by both the limits-to-growth exponents and the resource economics of the 1970s, population growth tends to be good for sustainability and economic growth; a possible counteracting factor, outside the framework considered here, might be that increased population density can generate congestion and aggravate environmental problems. (3) Whether or not there is population growth, endogenous technical change *may* bring about the technological basis for a rising per capita consumption in the long run or at least non-decreasing per capita consumption, but we can not be sure. (4) With diminishing returns to producible inputs, including knowledge, the long-run per capita growth rate is pinned down by growth in research effort. (5) Even when sustained growth is technologically feasible, if the rate of impatience is high enough, a utilitarian social planner's solution entails ultimately declining per capita consumption. (6) The standard approach to modelling endogenous technical change in a non-renewable resource set-up ignores that also R&D may need the resource (directly or indirectly). This biases the conclusions in an optimistic direction. Indeed, sustained per capita growth requires stronger parameter restrictions when the resource is "growth essential", than when it is not. (7) When the resource is "growth essential", then a policy aiming at stimulating long-run growth generally has to reduce the long-run depletion rate. In this sense promoting long-run growth and "supporting the environment" go hand in hand. This observation is of particular interest in view of the fact that changing the perspective from exogenous to endogenous technical progress means bringing a source of numerous market failures to light.

New growth theory has usually, as a simplifying device, considered population growth as exogenous. Given this premise, a key distinction – sometimes even controversy – arises between what is called fully endogenous growth and what is called semi-endogenous growth. In mainstream new growth theory, where non-renewable resources are completely left out of the analysis, this distinction tends to coincide with three other distinctions: (a) that between models that suffer from non-robustness due to a problematic knife-edge condition and models that do not; (b) that between models that imply a scale effect on growth and models that do not; and (c) models that imply policy-dependent long-run growth and models that do not. When non-renewable resources are taken into account and enter the growth engine (directly or indirectly), these dissimilarities are modified: (i) the non-robustness problem vanishes because of the disappearance of the critical knife-edge condition; yet, fully endogenous growth does not become more plausible than before, rather the contrary; (ii) the problem of a scale-effect on growth disappears; (iii) due to the presence of two very different assets, producible capital and non-producible resource deposits, even in the semi-endogenous growth case there is generally scope for policy having long-run growth effects.

The results listed here are, of course, subject to modification to the extent that non-renewable resources may not be essential in the long run. Similarly, a thorough integration of environmental aspects in the analysis deserves much more attention than this review has allowed.

Notes

* The activities of EPRU (Economic Policy Research Unit) are supported by a grant from The Danish National Research Foundation.

1. With a follow-up in Meadows et al. (1992).
2. There are signs that the current renewed rise in oil prices may have a similar effect.
3. The focus on *endogenous* technical change in a world with essential non-renewable resources differentiates this brief (and selective) review from other reviews of the economics of non-renewable resources (Solow 1974b, Dixit 1976, Dasgupta and Heal 1979, Withagen 1990, Heal 1998, and Krautkraemer 1998).
4. Whenever the term "growth" is used in this article, per capita growth is meant. For enlightening textbooks on new growth theory the reader is referred to Barro and Sala-i-Martin (2nd edn., 2004), Aghion and Howitt (1998) and, at a more elementary level, Jones (2002b). It should also be mentioned that new growth theory has important forerunners such as Nordhaus (1969) and Shell (1973).
5. Formaly, the growth engine of an endogenous growth model is defined as the set of input-producing sectors or activities using their own output as an input.
6. It is true that patents, concealment etc. can for a while exclude other firms from the commercial use of a specific innovation. Yet the general engineering principles behind the innovation are likely to diffuse rather quickly and add to the stock of common technical knowledge in society.
7. Synonymous with this is the sometimes used term *strictly endogenous growth*.
8. An analogue argument goes through for the vertical innovations models.
9. This knife-edge critique is equally relevant for the accumulation-based endogenous growth models (e.g. Lucas 1988 and Rebelo 1991), since they rely on a knife-edge condition similar to $\varphi = 1$.
10. The result that $g_y = g_A$ in a steady state follows, as a special case, by the method applied to balanced growth analysis in Section 3.3.
11. Emphasis on the non-rival character of technical knowledge is not specific to new growth theory, but can be found already in, e.g. Arrow (1962a) and Nordhaus (1969). What is new is rather the elaborate integration of this facet into dynamic general equilibrium models with imperfect competition.
12. This is usually ruled out by assumption. But not always. Indeed, one may allow for endogenous fertility, thereby endogenizing n (as in Jones 2003). And Cozzi (1997) develops a model where even φ is endogenous.
13. For more elaborate variants of the semi-endogenous approach, with detailed accounts of R&D and market structure, see Kortum (1997) and Segerstrom (1998). An early example is Arrow (1962b). A somewhat different way to aleviate or eliminate scale effects on growth is based on *adoption costs* (Jovanovic 1997).
14. Another strand is the new theories about how the market mechanism and profit incentives affect not only the *rate* of technical change, but also its *direction* (Acemoglu 2003). Yet a new strand, perhaps deserving to be categorized as *third-generation* models, is the integration of industrial organization theory and growth theory in an endeavour to achieve a nuanced understanding of the relationship between market structure and innovation (see, e.g. Aghion and Grifitt 2005).
15. At least within the second-generation framework this is so. To my knowledge there exists, so far, no compelling demonstration of fully endogenous growth arising generically from a more in-depth framework. Yet, Weitzman (1998a) is an attempt in this direction.
16. Not Charles I. Jones, but Larry E. Jones.
17. That is, marginal productivities are positive, but diminishing in own factor.
18. D-H-S-S had $\delta = 0$, thereby ignoring capital depreciation, because they considered exponential decay unrealistic and other depreciation formulas too cumbersome. Here, we allow $\delta > 0$, because exponential decay is a normal simplifying assumption in growth theory.

19. Thus the model's description of resource extraction is trivial. That is why it is natural to classify the model as a *one-sector* model notwithstanding there are two activities in the economy, manufacturing and resource extraction.
20. A more general case is $Y = \left[(1-\gamma)\tilde{F}(K,L)^{\psi} + \gamma R^{\psi}\right]^{1/\psi}$, where $\tilde{F}(K,L)$ has constant returns to scale. Here the elasticity of substitution between R and the "composite input" $\tilde{F}(K,L)$ is $1/(1-\psi)$, whereas that between K and L can be different (and may be variable). This makes it easier to obtain compliance with the empirical time trends in factor shares. A further generalization allows σ to depend on the input ratio $R/\tilde{F}(K,L)$. In fact, what really matters is whether $\sigma(R/\tilde{F}(K,L))$ remains low (below 1) for $R/\tilde{F}(K,L)$ approaching 0. Cass and Mitra (1991) generalize the D-H-S-S analysis by providing necessary and sufficient conditions for non-decreasing consumption in a capital-resource model with minimal technological restrictions, including allowance for extraction costs of many different kinds.
21. To avoid misunderstanding: by "Cobb-Douglas case" we refer to any function where R enters in a "Cobb-Douglas fashion", i.e. any function like $Y = \tilde{F}(K,L)^{1-\gamma}R^{\gamma}$.
22. Also the assumption $n = 0$ seems acceptable for the very long run on this finite planet. It appears harder to swallow $\delta = 0$, but a generalization of Solow's result is possible for certain patterns of non-exponential depreciation (Dasgupta and Heal 1979, p. 226).
23. For a historical account of energy technology, see Smil (1994).
24. For a lucid account of this theorem by Uzawa (1961), see Jones and Scrimgeour (2005).
25. For example, along *any* economic development path, the input of the non-renewable resource must in the long run asymptotically approach zero. From a physical point of view, however, there must be some minimum amount of the resource below which it cannot fulfill its role as a productive input. Thus, strictly speaking, sustainability requires that in the very long run non-renewable resources become inessential.
26. Suzuki (1976) has $\delta_A = \delta_K = 0$. But in order to comply with the general framework in this chapter, we allow $\delta_K > 0$, hence $\delta \geqslant 0$. Chiarella (1980) modifies (15) into $\dot{A} = I_A^{\xi}$, $\xi > 0$, and focuses on the resulting quite complicated transitional dynamics.
27. It is shown in Groth (2004) that "only if" in result (i) can be replaced by the stronger "if and only if". Note also that if some irreducibly exogenous element in the technological development is allowed in the model by replacing the constant B in equation (17) by $e^{\tau t}$, where $\tau \geqslant 0$, then equation (21) is replaced by $\tau + (\tilde{\alpha} + \beta - 1)n > 0$ or $\tilde{\alpha} > 1$. Both Stiglitz (1974a, p. 131) and Withagen (1990, p. 391) ignore implicitly the possibility $\tilde{\alpha} > 1$. Hence, from the outset they preclude fully endogenous growth.
28. See Groth (2004).
29. Suzuki's (1976) article also contains another model, with a resource externality. We touch upon this model in Section 6.
30. This is a reminder that the distinction between fully endogenous growth and semi-endogenous growth is not the same as the distinction between policy-dependent and policy-invariant growth.
31. If $\rho = n$, the improper integral U_0 tends to be unbounded and then the optimization criterion is not maximization, but "overtaking" or "catching-up" (see Seierstad and Sydsaeter 1987). For simplicity we have here ignored (as does Stiglitz) that also environmental quality should enter the utility function.
32. After Ramsey (1928).
33. After Hotelling (1931). Assuming perfect competition, the real resource price becomes $p_R = \partial Y/\partial R$ and the real rate of interest is $r = \partial Y/\partial K - \delta$. Then the rule takes the more familar form $\dot{p}_R/p_R = r$. If there are extraction costs at rate $C(R,S,t)$, then the rule takes the form $\dot{p}_S - \partial C/\partial S = rp_S$, where p_S is the price of the unextracted resource (whereas $p_R = p_S + \partial C/\partial R$).

    It is another thing that the rise in resource prices and the predicted decline in resource use have not yet shown up in the data (Krautkraemer 1998; Smil 2003); this may be due to better extraction technology and discovery of new deposits. But in the long run, if non-renewable resources *are* essential, this tendency inevitably will be reversed.

34. After the initial phase of complete specialization described in Section 4.1, we have, due to the proportionality between K, A, and $\tilde{K}$, that $\partial Y/\partial K = \partial Y/\partial A = \partial Y/\partial \tilde{K} = \tilde{\alpha} Y/\tilde{K}$. Notice that the Hotelling rule is independent of preferences; *any* path that is *efficient* must satisfy the Hotelling rule (as well as the exhaustion condition $\lim_{t\to\infty} S(t) = 0$).
35. As argued above, $\tilde{\alpha} < 1$ seems plausible. Generally, θ is estimated to be greater than one (see, e.g. Attanasio and Weber 1995); hence $D > 0$. The stability result as well as other findings reported here are documented in Groth and Schou (2002).
36. This aspect will become more lucid in the two-sector models of the next section, where the non-rival character of technical knowledge is more transparent.
37. This as well as the other results go through if a fixed resource like land is included as a necessary production factor. Indeed, letting J denote a fixed amount of land and replacing equation (14) by $Y = A^{\varepsilon} K^{\alpha} L^{\beta} R^{\gamma} J^{1-\alpha-\beta-\gamma}$, where now $\alpha + \beta + \gamma < 1$, leave equation (19)–(21), (26), and (27) unchanged.
38. Thus, if we do not require $D > 0$ in the first place, (iv) could be reformulated as: existence of a *stable* optimal BGP with $g_c > 0$ *requires* $n > 0$. This is not to say that reducing n from positive to zero renders an otherwise stable BGP instable. Stability-instability is governed solely by the sign of D. Given $D > 0$, letting n decrease from a level above the critical value, $\gamma\rho/(\tilde{\alpha} + \beta + \gamma - 1)$, given from equation (26), to a level below, changes g_c from positive to negative, i.e. growth comes to an end.
39. More commonplace observations are that increased impatience leads to faster depletion and lower growth (in the plausible case $\tilde{a} < 1$). Further, in the log-utility case ($\theta = 1$) the depletion rate u equals the effective rate of impatience, $\rho - n$.
40. Here we have in mind the fully endogenous growth literature. The results are more cognate with the results in semi-endogenous growth models without non-renewable resources, like Jones (1995b).
41. Essentialy this structure also characterizes the two-sector models by Robson (1980) and Takayama (1980), although these contributions do not fully comprehend the non-rival character of knowledge, since they have L_A/L in equation (29) instead of L_A.
42. Using patent data, Popp (2002) finds a strong, positive impact of energy prices on energy-saving innovations.
43. In this two-sector framework a BGP means a path along which Y, C, K, *and* N grow at constant rates (not necessarily positive). It is understood that the path considered is *efficient* and thus leaves nothing of the resource unutilized forever.
44. The first inequality ensures $u > 0$ (equivalent with the necessary transversality condition in the optimal control problem being satisfied), the second ensures $g_c > 0$.
45. For ease of interpretation we have written equation (38) on a form analogue to equation (37). In case $\varphi = 1$, equation (38) should be interpreted as $(1-\varphi)(1-\alpha)g_c = [\varepsilon\eta - (1-\varphi)\gamma]n - [\varepsilon(1-\eta) + (1-\varphi)\gamma]u$.
46. A possible reason for the popularity of the model of the previous section is that it has transitional dynamics that are less complicated than those of the present model (four-dimensional dynamics versus five-dimensional).
47. Although a scale effect on growth is absent, a positive scale effect on levels remains, as shown in Groth (2005). This is due to the non-rival character of technical knowledge.
48. As soon as $\varphi \geqslant 1$, growth becomes explosive.
49. For an exploration of this range, see Groth et al. (2006).

References

Acemoglu, D. (2003), ‘Labour- and Capital-augmenting Technical Change, *Journal of the European Economic Association* **1**, 1–37.

Aghion, P. and R. Griffith (2005), *Competition and Growth. Reconciling Theory and Evidence,* Cambridge MA: MIT Press.

Aghion, P. and P. Howitt (1992), 'A Model of Growth Through Creative Destruction', *Econometrica* **60**, 323–351.

Aghion, P. and P. Howitt (1998), *Endogenous Growth Theory*, Cambridge MA: MIT Press.

Alcalá, F. and A. Ciccone (2004), 'Trade and Productivity', *Quarterly Journal of Economics* **119**, 613–646.

André, F. J. and S. Smulders (2004), 'Energy Use, Endogenous Technical Change and Economic Growth', Working paper, Tilburg, The Netherlands.

Antweiler, W. and D. Trefler (2002), 'Increasing Returns and All That: A View from Trade', *American Economic Review* **92**, 93–119.

Arrow, K. J. (1962a), 'Economic Welfare and the Allocation of Resources for Invention', in *The Rate and Direction of Inventive Activity: Economic and Social Factors*, ed. by R. R. Nelson, Princeton University Press.

Arrow, K. J. (1962b), 'The Economic Implications of Learning by Doing', *Review of Economic Studies*, **29**, 153–73.

Attanasio, O. and G. Weber (1995), 'Is Consumption Growth Consistent with Intertemporal Optimization?' *Journal of Political Economy* **103**, 1121–1157.

Barro, R. J. and X. Sala-i-Martin (2004), *Economic Growth*, 2nd edn. Cambridge MA: MIT Press.

Bretschger, L. and S. Smulders (2006), 'Sustainability and Substitution of Exhaustible Natural Resources. How Resource Prices Affect Long-term R&D-investments, *Economics Working Paper Series 03/26*, rev. version May 2006, ETH Zürich.

Brock, W. A. and M. S. Taylor (2005), 'Economic Growth and the Environment: A Review of Theory and Empirics', in *Handbook of Economic Growth, Vol. 1B*, P. Aghion and S. N. Durlauf, eds. pp.1749–1821. Elsevier, Amsterdam.

Caballero, R. J. and R. K. Lyons (1992), The Case of External Economies, in *Political Economy, Growth, and Business Cycles*, A. Cukierman, Z. Hercowitz and L. Leiderman, eds. pp. 117–139, Cambridge MA: MIT Press.

Cass, D. and T. Mitra (1991), 'Indefinitely Sustained Consumption Despite Exhaustible Resources, *Economic Theory* **1** (2), 119–146.

Chiarella, C. (1980), 'Optimal Depletion of a Nonrenewable Resource when Technological Progress is Endogenous', in *Exhaustible Resources, Optimality, and Trade*, M. C. Kemp and N. V. Long, eds., North-Holland, Amsterdam.

Chichilnisky, G., G. M. Heal, and A. Vercelli, eds., (1998), *Sustainability: Dynamics and Uncertainty*. Kluwer, Dordrecht.

Cozzi, G. (1997), 'Exploring Growth Trajectories', *Journal of Economic Growth* **2**, 385–398.

Dasgupta, P. (1993), Natural Resources in an Age of Substitutability', in A. V. Kneese and J. L. Sweeney, eds., *Handbook of Natural Resource and Energy Economics*, Vol. 3, pp. 1111–1130. Elsevier, Amsterdam.

Dasgupta, P. and G. M. Heal (1974), 'The Optimal Depletion of Exhaustible Resources', *Review of Economic Studies* **41**, Symposium, 3–28.

Dasgupta, P., and G. M. Heal (1979), *Economic Theory and Exhaustible Resources*. Cambridge: Cambridge University Press.

Dasgupta, P., G. M. Heal, and M. K. Majumdar (1977), 'Resource Depletion and Research and Development', in M. D. Intrilligator ed., *Frontiers of Quantitative Economics, Vol. IIIB*, 483–505, North-Holland, Amsterdam.

Dasgupta, P., G. M. Heal, and A. K. Pand (1980), Funding Research and Development, *Applied Mathematical Modeling* **4**, 87–94.

Di Maria, C. and S. Valente (2006), 'The Direction of Technical Change in Capital-resource Economies, *Economics Working Paper Series 06/50*, ETH Zürich.

Dinopoulos, E. and P. Thompson (1998), 'Schumpeterian Growth without Scale Effects', *Journal of Economic Growth* **3**, 313–335.

Dixit, A. K. (1976), *The Theory of Equilibrium Growth*, Oxford: Oxford University Press.

Evans, P. (1996), Using Cross-country Variances to Evaluate Growth Theories', *Journal of Economic Dynamics and Control* **20**, 1027–1049.

Farzin, Y. H. and O. Tahvonen (1996), 'Global Carbon Cycle and the Optimal Time Path of a Carbon Tax, *Oxford Economic Papers* **48**, 515–536.

Fullerton, D. and S.-R. Kim (2006), 'Environmental Investment and Policy with Distortionary Taxes and Endogenous Growth', *NBER WP 12070.*

Grimaud, A. and L. Rougé (2003), 'Non-renewable Resources and Growth with Vertical Innovations: Optimum, Equilibrium and Economic Policies, *Journal of Environmental Economics and Management* **45**, 433–453.

Grossman, G. M. and E. Helpman (1991), Innovation and Growth in the Global Economy, Cambridge MA: MIT Press.

Groth, C. (2004), 'Strictly Endogenous Growth with Non-renewable Resources Implies an Unbounded Growth Rate, *Topics in Macroeconomics* **4** (1), 1–13.

Groth, C. (2005), 'Growth and Non-renewable Resources Revisited, *Working paper*, University of Copenhagen.

Groth, C. (2006), 'Capital and Resource Taxation in a Romer-style Growth Model with Non-renewable Resources, *Working paper*, University of Copenhagen.

Groth, C. and P. Schou (2002), 'Can Nonrenewable Resources Alleviate the Knife-edge Character of Endogenous Growth?", *Oxford Economic Papers* **54**, 386–411.

Groth, C., and P. Schou (2006), 'Growth and Non-renewable Resources: The Different Roles of Capital and Resource Taxes, *Journal of Environmental Economics and Management*, forthcoming.

Groth, C., K.-J. Koch, and T. Steger (2006), 'Rethinking the Concept of Long-run Economic Growth', *CESifo Working Paper no. 1701*, Munich.

Hall, R. E. (1990), 'Invariance Properties of Solow's Residual', in P. Diamond, ed., *Growth/Productivity/Un-employment. Essays to Celebrate Bob Solow's Birthday*, pp. 71–112. Cambridge MA: MIT Press.

Hartwick, J. M. (1977), Intergenerational Equity and the Investing of Rents from Exhaustible Resources, *American Economic Review* **67**, 972–974.

Heal, G. M. (1998), *Valuing the Future. Economic Theory and Sustainability.* New York. Columbia University Press.

Hotelling, H. (1931), 'The Economics of Exhaustible Resources', *Journal of Political Economy* **39**, 137–175.

Howitt, P. (1999), 'Steady Endogenous Growth with Population and R&D Inputs Growing', *Journal of Political Economy* **107**, 715–730.

Jones, C. I. (1995a), 'Time-series Tests of Endogenous Growth Models', *Quarterly Journal of Economics* **110**, 495–525.

Jones, C. I. (1995b), 'R&D-based Models of Economic Growth', *Journal of Political Economy* **103**, 759–784.

Jones, C. I. (1999), 'Growth: With or without scale effects?', *American Economic Review* **89**, May, 139–144.

Jones, C. I. (2002a), 'Sources of U.S. Economic Growth in a World of Ideas', *American Economic Review* **92**, 220–239.

Jones, C. I. (2002b), *Introduction to Economic Growth*, 2nd edn. New York: W.W. Norton.

Jones, C. I. (2003), 'Population and Ideas: A theory of Endogenous Growth, in P. Aghion, R. Frydman, J. Stiglitz, and M. Woodford, *Knowledge, Information, and Expectations in Modern Macroeconomics: In Honor of Edmund S. Phelps*, pp. 498–521, Princeton: Princeton University Press.

Jones, C. I. (2005), 'The Shape of Production Functions and the Direction of Technical Change', *Quarterly Journal of Economics* **120** (2), 517–549.

Jones, C. I. and D. Scrimgeour (2005), 'The Steady-state Growth Theorem: Understanding Uzawa (1961)', *working paper*, Berkeley, University of California.

Jones, L. E. and R. E. Manuelli (1997), 'The Sources of Growth', *Journal of Economic Dynamics and Control* **21**, 75–114.

Jovanovic, B. (1997), 'Learning and Growth', in D. M. Kreps and K. F. Wallis, eds., *Advances in Economics and Econometrics: Theory and Applications, Vol. II*, pp.318–339, Cambridge: Cambridge University Press.

Just, R. E., S. Netanyahu, and L. J. Olsen (2005), 'Depletion of Natural Resources, Technological Uncertainty, and the Adoption of Technological Substitutes', *Resource and Energy Economics* **27**, 91–108.

Kamien, M. I. and N. L. Schwartz (1978), 'Optimal Exhaustible Resource Depletion with Endogenous Technical Change', *Review of Economic Studies* **45** (1), 179–196.

Kortum, S. (1997), 'Research, Patenting, and Technological Change', *Econometrica* **65**, 1389–1419.

Krautkraemer, J. (1985), 'Optimal Growth, Resource Amenities and the Preservation of Natural Environments', *Review of Economic Studies* **52**, 153–170.

Krautkraemer, J. (1998), 'Nonrenewable Resource Scarcity', *Journal of Economic Literature* **36** (4), 2065–2107.

Kremer, M. (1993), Population Growth and Technological Change: One million B.C. to 1990, *Quarterly Journal of Economics* **108**, 681–716.

Li, C.-W. (2000), 'Endogenous vs. Semi-endogenous Growth in a Two-sector Model', *Economic Journal* **110**, C109–C122.

Li, C.-W. (2002a), 'Growth and Scale Effects: The Role of Knowledge Spillovers', *Economics Letters* **74**, 177–185.

Li, D. (2002b), 'Is the AK Model Still Alive? The Long-run Relation Between Growth and Investment Re-examined, *Canadian Journal of Economics* **35** (1), 92–114.

Lucas, R. E., Jr. (1988), 'On the Mechanics of Economic Development, *Journal of Monetary Economics* **22**, 3–42.

Mankiw, N. G. (1995), 'The Growth of Nations', *Brookings Papers on Economic Activity*, (1), 275–326.

Meadows, D. H. et al. (1972), *The Limits to Growth*. New York: Universe Books.

Meadows, D. H., D. L. Meadows, and J. Randers (1992), *Beyond the Limits*. Post Mills: Chelsea Green Publ. Com.

Neumayer, E. (2000), 'Scarce or Abundant? The Economics of Natural Resource Availability', *Journal of Economic Surveys* **14** (3), 307–335.

Nordhaus, W. (1969), 'An Economic Theory of Technological Change', *American Economic Review* **59** (2), 18–28.

Nordhaus, W. (1992), 'Lethal Model 2: The Limits to Growth Revisited', *Brookings Papers on Economic Activities*, (2), 1–43.

Nordhaus, W. and J. Tobin (1972), 'Is growth obsolete?' in: Economic Growth: Fiftieth Anniversary Colloquium V. New York: Columbia University Press. Reprinted in J. Tobin, *Essays in Economics. Theory and Policy* 1985, pp.360–450. Cambridge, MA: MIT Press.

Peretto, P. (1998), 'Technological Change and Population Growth', *Journal of Economic Growth* **3** (4), 283–311.

Peretto, P. and S. Smulders (2002), Technological Distance, Growth and Scale Effects, *Economic Journal* **112**, 603–624.

Popp, D. (2002), Induced Innovation and Energy Prices, *American Economic Review* **92** (1), 160–180.

Ramsey, F. (1928), 'A Mathematical Theory of Saving', *Economic Journal* **38**, 543–559.

Rebelo, S. (1991), 'Long-run Policy Analysis and Long-run Growth', *Journal of Political Economy* **99**, 500–521.

Robson, A. J. (1980), 'Costly Innovation and Natural Resources', *International Economic Review* **21**, 17–30.

Romer, P. M. (1986), 'Increasing Returns and Long-run Growth', *Journal of Political Economy* **94**, 1002–1037.

Romer, P. M. (1987), 'Growth Based on Increasing Returns due to Specialization', *American Economic Review* **77**, May, 56–62.

Romer, P. M. (1990), 'Endogenous Technical Change', *Journal of Political Economy* **98** (Suppl.), 71–102.

Romer, P. M. (1995), 'Comment to Srinivasan', in T. Ito and A. O. Krueger, eds., *Growth Theories in the Light of the East Asian Experience*, pp. 66–70, Chicago IL: University of Chicago Press.

Romero-Avila, D. (2006), 'Can the AK Model be Rescued? New Evidence from Unit Root Tests with Good Size and Power', *Topics in Macroeconomics* **6** (1), 0–40.

Scholz, C. M. and G. Ziemes (1999) 'Exhaustible Resources, Monopolistic Competition, and Endogenous Growth', *Environmental and Resource Economics* **13**, 169–185.

Schou, P. (2000), 'Polluting Nonrenewable Resources and Growth', *Environmental and Resource Economics* **16**, 211–27.

Schou, P. (2002), 'When Environmental Policy is Superfluous: Growth and Polluting Resources', Scandinavian Journal of Economics **104**, 605–620.

Segerstrom, P. (1998), 'Endogenous Growth Without Scale Effects', *American Economic Review* **88**, 1290–1310.

Seierstad, A. and K. Sydsaeter (1987), *Optimal Control Theory with Economic Applications*, Amsterdam: North-Holland.

Shell, K. (1973), 'Inventive Activity, Industrial Organization and Economic Growth', in J. A. Mirrlees and N. H. Stern, eds., *Models of Economic Growth*. pp. 77–96, London: Macmillan.

Sinclair, P. (1994), On the Optimum Trend of Fossil Fuel Taxation, *Oxford Economic Papers* **46**, 869–877.

Smil, V. (1994), *Energy in World History*. Boulder Co: Westview Press.

Smil, V. (2003), *Energy at the Crossroads. Global Perspectives and Uncertainties*. Cambridge MA: MIT Press.

Smulders, S. and M. de Nooij (2003), 'The Impact of Energy Conservation on Technology and Economic Growth', *Resource and Energy Economics* **25**, 59–79.

Solow, R. M. (1974a), 'Intergenerational Equity and Exhaustible Resources', *Review of Economic Studies* **41**, Symposium Issue, 29–45.

Solow, R. M. (1974b), 'The Economics of Resources or the Resources of Economics', *American Economic Review Papers and Proceedings*, **64**, 1–14.

Solow, R. M. (1994), 'Perspectives on Growth Theory', *Journal of Economic Perspectives* **8**, 45–54.

Stiglitz, J. (1974a), Growth with Exhaustible Natural Resources: Efficient and Optimal Growth paths, *Review of Economic Studies Symposium Issue*, **41**, 123–137.

Stiglitz, J. (1974b), 'Growth with Exhaustible Natural Resources: The Competitive Economy', *Review of Economic Studies Symposium Issue*, **41**, 139–152.

Suzuki, H. (1976), On the Possibility of Steadily Growing Per Capita Consumption in an Economy With a Wasting and Non-replenishable Resource', *Review of Economic Studies* **43**, 527–535.

Tahvonen, O. and S. Salo (2001), 'Economic Growth and Transitions between Renewable and Non-renewable Energy Resources', *European Economic Review* **45**, 1379–1398.

Takayama, A. (1980), 'Optimal Technical Progress with Exhaustible Resources', in M. C. Kemp and N. V. Long, eds., *Exhaustible Resources, Optimality, and Trade*, pp.95–110, Amsterdam: North-Holland.

United Nations (2005), *World Population Prospects*. The 2004 Revision, Population Division, New York.

Uzawa, H. (1961), 'Neutral Inventions and the Stability of Growth Equilibrium', *Review of Economic Studies* **28** (2), 117–124.

Young, A. (1998), 'Growth Without Scale Effects', *American Economic Review* **88**, 41–63.

Weitzman, M. L. (1998a), 'Recombinant Growth', *Quarterly Journal of Economics* **113**, 331–360.

Weitzman, M. L. (1998b), 'Why the Far-distant Future Should be Discounted at its Lowest Possible Rate', *Journal of Environmental Economics and Management* **36**, 201–208.

Weitzman, M. L. (2001), 'Gamma Discounting', *American Economic Review* **91** (1), 260–271.

Withagen, C. (1990), 'Topics in Resource Economics', in F. van der Ploeg, ed., *Advanced Lectures in Quantitative Economics*, pp. 381–419. London: Academic Press.

8. Sectoral Energy- and Labour-Productivity Convergence

PETER MULDER[1,2] AND HENRI L.F. DE GROOT[2,3,*]

[1]*Gabinete de Estudos, Ministry of Planning and Development, Maputo, Mozambique;*
[2]*Department of Spatial Economics, Vrije Universiteit, De Boelelaan 1105, 1081 HV Amsterdam The Netherlands;*
[3]*CPB Netherlands Bureau for Economic Policy Analysis, The Hague, The Netherlands;*
[*]*Author for correspondence (e-mail: hgroot@feweb.vu.nl)*

This paper empirically investigates the development of cross-country differences in energy- and labour productivity. The analysis is performed at a detailed sectoral level for 14 OECD countries, covering the period 1970–1997. A σ-convergence analysis reveals that the development over time of the cross-country variation in productivity performance differs across sectors as well as across different levels of aggregation. Both patterns of convergence as well as divergence are found. Cross-country variation of productivity levels is typically larger for energy than for labour. A β-convergence analysis provides support for the hypothesis that in most sectors lagging countries tend to catch up with technological leaders, in particular in terms of energy productivity. Moreover, the results show that convergence is conditional, meaning that productivity levels converge to country-specific steady states. Energy prices and wages are shown to positively affect energy- and labour-productivity growth, respectively. We also find evidence of economies of scale, whereas the investment share, openness and specialization play only a modest role in explaining cross-country variation in energy- and labour-productivity growth.

Key words: convergence, energy productivity, labour productivity, sectoral analysis

1. Introduction

Over the last decades increasing attention has been paid to the role of energy in production processes and to its importance for economic growth. Energy consumption is, however, also an important source of greenhouse gas emissions. Most governments in OECD countries explicitly recognize the need for sustainable development and aim at a decoupling of economic growth and environmental pressure. In a more operational sense, this underlines the importance of sustained growth of both labour and energy productivity. Productivity growth is thought to be determined not only by country-specific characteristics, such as investments and factor prices, but also by developments in the outside world. Therefore, an important issue in understanding long-run productivity performance is whether the process of economic growth tends to involve reductions in productivity differences among countries, for example, due to diminishing returns to capital accumulation

L. Bretschger and S. Smulders (eds.), Sustainable Resource Use and Economic Dynamics, 165–190.

or technology transfers. In this paper we explore differences in energy productivity across countries and across sectors, and compare them with differences in labour productivity. Are these differences decreasing, or is the gap between leading and backward countries getting larger? Are patterns of energy-productivity similar to those of labour-productivity convergence? Do relatively inefficient countries catch up with technological 'leaders' in a globalizing world? And if so, how quickly and by what means? We aim to answer these questions by simultaneously carrying out an empirical analysis of cross-country energy- and labour-productivity convergence at a detailed sector level, using a new dataset that merges energy data and economic data for 13 sectors and 14 OECD countries, covering the period 1970–1997.

In several respects, our paper differs from previous empirical research on cross-country productivity convergence. It extends the empirical macroeconomic convergence literature to energy-productivity developments (see also Miketa and Mulder 2005, for a complementary paper[1]). In spite of many existing cross-country studies on energy-productivity or energy-intensity developments and its determinants (for example Howarth et al. 1991; Miketa 2001; Schipper and Meyers 1992; Unander et al. 1999; Mulder and de Groot 2003a), systematic analyses of convergence from a macroeconomic perspective are rare. Hence, we add to the existing literature a systematic comparison of energy- and labour-productivity convergence, whereby the latter mainly serves as a point of reference for our analysis of energy-productivity convergence. Furthermore, we do so at a detailed sectoral level. By looking at cross-country convergence patterns *within* sectors, our analysis differs from virtually all convergence studies in the empirical growth literature, since they employ aggregate data. Important exceptions are sectoral studies by Dollar and Wolff (1988, 1993) and Bernard and Jones (1996a, b) who – using (partly) the same data source as we do (OECD's ISDB) – conclude that a convergence analysis of aggregate productivity levels masks substantial differences at the sectoral level. An important underlying reason for this result is that productivity levels, measured as the ratio of value added over a unit of input (viz. energy and labour), can substantially differ among sectors because some activities require inherently more capital, higher labour skills and/or technology than others. Aggregate productivity trends are therefore not directly attributable to technological change in individual sectors, as they can also be the result of changes in the distribution of production factors among sectors. Our sectoral approach corrects for most of the impact of such changes in the structure of production on aggregate productivity developments and, hence, establishes a closer link to issues concerning international convergence of technology-driven productivity performance. Our analysis differs from the previously mentioned sectoral convergence analyses in comparing labour- and energy-productivity convergence, in further disaggregating the manufacturing sector into 10 sub-sectors,[2] in using more recent data and in carrying out a more extensive search for country- and sector-specific factors to explain productivity convergence patterns.

The paper proceeds as follows. Section 2 discusses the theoretical background for this paper including the different notions of convergence that are found in the literature. Section 3 describes the data used. In Section 4 we analyze the development of cross-country differences of energy- and labour-productivity levels within sectors over time. In Section 5 we use a panel-data approach to test the proposition that sectoral growth rates of energy and labour productivity are inversely related to the initial levels of energy and labour productivity, indicating possible patterns of catching-up. In addition, we try to identify the country- and sector-specific fundamentals determining (differences in) energy- and labour productivity developments. Section 6 summarises and concludes.

2. Theoretical Background

The concept of productivity convergence has its roots in neoclassical growth theory, with its central notion of a transitional growth path to a steady state. The evolution of the literature on economic growth and productivity developments has resulted in a broad consensus on which are the key factors driving productivity growth across countries, and thus determining patterns of convergence and divergence. To structure a brief discussion of this issue and illuminate the various factors and mechanisms that may affect cross-country energy- and labour-productivity differences, let us take a neoclassical Cobb-Douglas production function:

$$Y = AK^{\alpha}L^{\beta}E^{1-\alpha-\beta} \tag{1}$$

where Y is output, A is technology, K is capital, L is labour and E is energy. Assuming that each input is paid according to its marginal product, equation (1) can be rewritten in terms of average energy- and labour-productivity as follows:

$$\frac{Y}{E} = A\left(\frac{K}{E}\right)^{\alpha}\left(\frac{L}{E}\right)^{\beta} = A\left(\frac{\alpha p_E}{(1-\alpha-\beta)r}\right)^{\alpha}\left(\frac{\beta p_E}{(1-\alpha-\beta)w}\right)^{\beta} \tag{2a}$$

$$\frac{Y}{L} = A\left(\frac{K}{E}\right)^{\alpha}\left(\frac{L}{E}\right)^{\beta-1} = A\left(\frac{\alpha p_E}{(1-\alpha-\beta)r}\right)^{\alpha}\left(\frac{\beta p_E}{(1-\alpha-\beta)w}\right)^{\beta-1} \tag{2b}$$

with r, w and p_E representing, respectively, the rental price of capital, the wage rate and the energy price. From these equations it can be easily seen that cross-country differences in energy and labour productivity may arise from differences in factor input ratios, (relative) factor prices and the level of technological development. Concerning the dynamics, these differences may change over time as the result of factor accumulation, factor price changes and technological change, which in turn can be facilitated by processes such as trade, foreign direct investment (FDI), learning and market conditions. Thus, these phenomena are among the key factors causing cross-country productivity differences to change over time, leading to patterns of convergence or divergence.

Returning to neoclassical growth theory, the Solow-Swan model (Solow 1956; Swan 1956) postulates convergence of per capita income, driven by the assumption of diminishing returns to capital accumulation at the economy-wide level. The dynamics of the model imply that initial differences in per capita income and capital endowments vanish in the long run. In the steady state, diminishing returns are offset by technological progress, the principal source of long-run economic growth. The new or endogenous growth theory (Lucas 1988; Romer 1986, 1990) yields a more diverse picture concerning patterns of convergence. It builds on the notion that capital should be considered as a broad concept, including human and intangible capital. In this theory, economic growth is driven by accumulation of knowledge or human capital, which is (at least partially) a public good. Hence, cross-country convergence depends on the extent of international knowledge spill-overs, allowing less productive countries to catch up with more advanced economies. As such, endogenous growth theory supports the old hypothesis of the existence of an 'advantage of backwardness' (Gerschenkron 1952), suggesting that being backward in productivity carries a potential for rapid advance (see, e.g., Abramovitz, 1986). At the same time, endogenous growth theory suggests that growth differentials may persist or even increase: learning effects, externalities and market imperfections allow for economy-wide increasing returns to capital accumulation and the existence of multiple steady-states. Moreover, there is some reason to believe that technology diffusion and knowledge spillovers are local rather than global (see, for example, Keller 2002) which raises the possibility that convergence patterns depend on the spatial dimension of technological progression.

A mixed view on convergence patterns also emerges if one takes into account the role of international trade and FDI. Trade and FDI could enhance cross-country convergence through knowledge diffusion and thus diminishing cross-country differences in technology level, and through convergence in factor prices via increasing international competition. On the other hand, trade and FDI could contribute to cross-country divergence by stimulating differential factor accumulation across countries, for example, because trade advances international specialization (Grossman and Helpman 1991). These various approaches generated some degree of controversy around the issue of convergence and caused the convergence hypothesis to be the subject of extensive empirical research.[3] In this paper we do not go further into this debate, but focus instead on the cross-country differences in energy productivity, whereas the empirical convergence literature focuses on convergence of per capita income, labour productivity and total factor productivity.

Our focus on the development of cross-country energy-productivity requires some additional discussion on the driving forces behind the evolution of differences in energy productivity across countries. In a recently developed augmented version of the Solow model including pollution (emissions) as a joint product of output as well as technological progress in abatement, Brock and Taylor (2004) show that one can expect cross-country convergence of emissions per capita as nations get richer. They argue this to be in line with the empirical evidence on the existence of an Environmental Kuznets Curve. This pattern of emission intensity convergence is caused by the joint forces of output convergence and technology catch-up, possibly enhanced by sectoral shifts away from heavy industry as well as reduced cross-country heterogeneity (for example in terms of population growth and savings rates). But, contrary to emissions, energy is an intermediate input into the production process rather than a joint product of output. This implies that unless energy is strongly complementary to capital, there is not much reason to believe that diminishing returns to capital (in some augmented version of the Solow model including energy) are an important source of energy-productivity convergence across countries.[4] However, from classical as well as recent empirical research, it is well known that technological change, prices and changes in economic structure (sectoral shifts) are key determinants of aggregate energy-productivity growth (see, for example, Berndt 1978; Jorgenson 1984; Schipper and Meyers 1992).

As noted above, knowledge spillovers and international technology diffusion – possibly facilitated by (increasing) international trade and FDI – may lead to processes of technology catch-up which in turn can be a potentially important source of energy-productivity convergence. In any case, since productivity growth is primarily driven by technological change, the evidence of conditional convergence reported in this paper suggests that patterns of international energy-saving technology flows exist, while at the same time they seem to be limited and at least to some extent sector-specific. Obviously, decreasing cross-county differences in energy taxation affect energy productivity through the (relative) final prices of the input factor energy, and depend, among others, on (cross-country differences in) government policies, institutions, openness, trade and market conditions. The role of (differences in) economic structure on aggregate energy productivity growth can be easily seen if we decompose the ratio of output to energy according to:

$$\frac{Y}{E} = \sum_{s=1}^{S} \frac{Y_s}{E_s} \frac{E_s}{E_T} \tag{3}$$

with s denoting the sectors of the economy, S indicating the total number of all sectors considered and E_T representing total final energy consumption. So, equation (3) says that aggregate energy productivity is the sum of the energy productivity of each sub-sector (the first term on the RHS) multiplied by the energy share of each sub-sector (the second term on the RHS). The first term on the RHS is sometimes referred to as the structural effect, and indicates the effect of changes in the structure of production on aggregate productivity growth. With shifts away from (heavy) industry to services as countries get richer, the resulting decreasing cross-country heterogeneity in the economy's structure may lead to decreasing cross-country differences in aggregate energy productivity. Of course, a similar argument applies to labour productivity. Finally, several authors have stressed the fact that changes in energy mix are an important source of aggregate energy productivity developments, because some energy types (such as natural gas and electricity) are more efficient than others (such as coal and oil) in terms of available energy (see, for example, Berndt 1978; Cleveland et al. 2000; Kaufmann 2004). This can be illustrated with equation (3), if we assume s now to represent various energy types and S the total number of different energy types used in the economy, with E_T representing total final energy consumption (expressed in a uniform unit such as ktoe). Then aggregate energy productivity is the product of the relative efficiency of the different energy types used (the first term on the RHS) and the relative shares of these different energy types (the second term on the RHS). Hence a decreasing cross-country heterogeneity in the use of various energy types is expected to contribute to cross-country convergence of aggregate energy productivity.[5]

Apart from the factors that may give rise to energy-productivity convergence, it is important to assess whether countries converge to a global or a local steady state. To address this question, the concept of conditional as opposed to unconditional convergence has been developed in the literature on convergence. The former concept posits that countries all converge to their own steady state whereas the latter assumes the existence of one steady state that is common to all countries. The empirical growth literature has found strong support for the notion of conditional convergence or club convergence (see, for example, Durlauf and Johnson 1992, Chatterji 1992, and Quah 1997 for seminal contributions). From this literature it follows that convergence can be understood in terms of levels and growth rates, which translates into a distinction between so-called σ-convergence and β-convergence (for example, Barro 1991, Barro and Sala-i-Martin 1992). The former refers to a decreasing variation of cross-country differences in productivity levels, while the latter suggests a tendency of countries with relatively low initial productivity levels to grow relatively fast, building upon the proposition that growth rates tend to decline as countries approach their steady state.[6] In this paper we will explore both patterns of σ-convergence and β-convergence. Moreover, we test whether energy and labour productivity convergence is conditional or unconditional and which are the factors explaining (cross country differences in) labour and energy productivity growth.

3. Data

The analysis presented in this paper is based on a newly constructed database that merges energy data from the Energy Balances as they are published by the International Energy Agency (IEA), and economic data from the International Sectoral Database (ISDB) and the Structural Analysis Database (STAN), both published by the OECD.[7] The main idea behind the construction of this database is to establish a link between economic and energy data at a detailed sectoral level. This results in the sector classification as described in Table I.

The database covers the period 1970–1997 and includes the following countries: Australia, Belgium, Canada, Denmark, Finland, France, West-Germany, Italy, Japan, the Netherlands, Norway, Sweden, United Kingdom and the United States.

Table I. Sector Classification

	Sector	Abbreviation	ISIC Rev. 2 code
1	Food and Tobacco	FOD	31
2	Textiles and Leather	TEX	32
3	Wood and Wood Products	WOD	331[a]
4	Paper, Pulp and Printing	PAP	34
5	Chemicals	CHE	351 + 352[b]
6	Non-Metallic Minerals	NMM	36
7	Iron and Steel	IAS	371
8	Non-Ferrous Metals	NFM	372
9	Machinery	MAC	381 + 382 + 383[c]
10	Transport Equipment	MTR	384
11	Construction	CST	50
12	Services	SRV	61 + 62 + 63 + 72 + 81 + 82 + 83 + 90[d]
13	Transport	TAS	71
14	Agriculture	AGR	10

[a]WOD excludes furniture since the sector WOD in the IEA Energy Balances excludes furniture.
[b]CHE includes non-energetic energy consumption, i.e. using energy carriers as feedstock.
[c]MAC = Metal Products (BMA, 381) + Agricultural and Industrial Machinery (MAI, 382) + Electrical Goods (MEL, 383).
[d]SRV = Wholesale and retail trade, restaurants and hotels (RET) + Communication (COM) + Finance, insurance, real estate and business services (FNI) + Community, social and personal services (SOC).

We measure energy productivity by gross value added per unit of final energy consumption and labour productivity by gross value added per worker (in full time equivalents). Value added is the net economic output of a sector, measured by the price differential between the price of output and the cost of input and comprises compensation to employees, operating surplus, the consumption of fixed capital and the excess of indirect taxes over subsidies (OECD 1998). Following the IEA, energy use is defined as final energy consumption in kilo tonnes of oil equivalence (ktoe), with sectoral data excluding transformation losses. Total employment is measured in the full-time equivalent number of persons, including self-employed.

Moreover, the database includes data on Investment, Energy Prices, Compensation of Employees, Export and Import – all at the sectoral level. The sector-specific energy prices are constructed by dividing sector-specific expenditures on energy over total sectoral energy consumption. The sector-specific expenditures are calculated as the product of the sectoral consumption of the four main energy carriers (Coal, Natural Gas, Electricity, Oil) – available from the Energy Balances – and the (annual) price of each energy carrier at the aggregate industrial sector – available from the IEA Energy Prices and Taxes series. In addition, some missing aggregate energy price data series have been constructed (see the Annex to this paper for details). Detailed descriptive statistics per sector and per country covering the growth rate of energy- and labour productivity, the log-levels of energy- and labour productivity and of GDP, and the levels of wages, energy prices, investment shares, openness, the Balassa indices and sector shares can be found in the Annex to this paper. The latter four variables are introduced and discussed in Section 5.2.

All currency-denominated variables are in 1990 US$ and have been converted by the OECD using 1990 purchasing power parities (PPPs). In principle, the theoretically most appropriate conversion factors for productivity comparisons at the sectoral level are to be based on a comparison of output prices by industry of origin, rather than on expenditure prices (see, for example, van Ark and Pilat 1993). Expenditure PPPs exclude the part of output that is exported, while they include imported goods produced elsewhere; they take account of differences in trade and transport margins and indirect taxes between countries, and they do not cover intermediate products. The main problem in using the production or industry-of-origin approach, however, is the limited availability of producer-price based PPPs, in particular for non-Manufacturing sectors (van Ark 1993). Moreover, we have no a priori reason to presume that the drawbacks of expenditure PPPs differ substantially across countries. Hence, we follow most studies in using expenditure PPPs. This enables us to do a systematic cross-country convergence analysis of energy- and labour-productivity performance at a high level of sectoral detail. Obviously, the results presented in this paper should be interpreted with caution, bearing in mind the before mentioned issue (see Sørensen 2001, and Bernard and Jones 2001 for a discussion).

4. σ-Convergence

This section deals with the notion of convergence in terms of levels. Do cross-country differences in energy- and labour-productivity levels decrease over time? Are patterns of energy-productivity convergence similar to those of labour-productivity convergence? And to what extent do the results depend on the level of aggregation? To answer these questions we calculated for each (sub-)sector – based on a balanced sample of 14 OECD countries (insofar as data are available) – the yearly unweighted cross-country standard deviation (σ) of the log of energy and labour productivity.[8] Table II shows the results for the years 1976 and 1990. Results for the entire time span for which data are available are graphically presented in the Annex to this paper. None of the results described in the remainder are peculiar to the choice of the two years for which the standard deviation is presented in Table II.

The macroeconomic development of the standard deviation of the log of 'energy- and labour-productivity levels (with 'macroeconomic' referring to the sum of aggregate Manufacturing, Transport, Services and Agriculture) reveals that cross-country differences in energy-productivity levels are substantially larger than cross-country differences of labour-productivity levels. Moreover, it can be seen that over time the standard deviation of the log of energy-productivity performance is increasing, indicating σ-divergence, while the opposite is true for cross-country labour-productivity performance, displaying a pattern of σ-convergence.

Table II. Standard deviation of log of energy-and labour productivity, 1976 and 1990

	Energy productivity		Labour productivity	
	1976	1990	1976	1990
Macroeconomic level[a]	0.261	0.294	0.210	0.171
Main sectors				
Manufacturing[b]	0.444	0.512	0.212	0.204
Services[c]	0.839	0.605	0.220	0.172
Transport[d]	0.510	0.439	0.278	0.248
Agriculture[b]	0.492	0.320	0.305	0.256
Manufacturing sectors				
Chemicals[g]	0.519	0.557	0.366	0.265
Food and Tobacco[i]	0.546	0.436	0.267	0.258
Iron and Steel[e]	0.468	0.580	0.481	0.278
Machinery[h]	0.570	0.350	0.202	0.239
Transport Equipment[i]	0.473	0.401	0.248	0.241
Non-Ferrous Metals[f]	0.473	0.660	0.426	0.313
Non-Metallic Minerals[g]	0.467	0.269	0.226	0.187
Paper, Pulp and Printing[e]	0.934	0.950	0.252	0.176
Textiles and Leather[i]	0.359	0.300	0.203	0.190
Wood and Wood Products[j]	0.887	0.848	0.362	0.225

[a]Excludes Canada, Japan, The Netherlands and Sweden due to limited data availability.
[b]Excludes Japan and The Netherlands due to limited data availability.
[c]Excludes The Netherlands and Sweden due to limited data availability.
[d]Excludes Canada and The Netherlands due to limited data availability.
[e]Excludes Australia and Japan due to limited data availability.
[f]Excludes Australia and Denmark due to limited data availability.
[g]Excludes Australia due to limited data availability.
[h]Excludes Australia, Canada, Japan and The Netherlands due to limited data availability.
[i]Excludes Australia and Canada due to limited data availability.
[j]Excludes Australia, Canada, France, Japan, United Kingdom and USA due to limited data availability.

As we noted in the introduction, a convergence analysis at aggregate levels may mask considerable variation in sectoral productivity developments (cf. Bernard and Jones 1996a, b; Dollar and Wolff 1988, 1993). Therefore, we continue by examining the development of cross-country productivity differentials within different sectors, viz. (aggregate) Manufacturing, Transport, Services and Agriculture. It can clearly be seen that only Manufacturing resembles the macroeconomic pattern of σ-divergence for energy productivity. Transport, Agriculture, and, in particular, Services, display evidence of σ-convergence. Note that the cross-country variation is relatively high in Services, which is to a large extent due to the exceptional and so far unexplained energy-productivity performance of Finland and Italy.[9] The macro-economic pattern of σ-convergence for labour productivity is only evident in Services and to a lesser extent in the Agricultural sector. Variation in cross-country productivity differentials remains overall fairly constant within aggregate Manufacturing and Transport. Comparing the results for energy and labour productivity reveals again that in each sector the cross-country variation of energy productivity is substantially larger than of labour productivity. They accord well with the findings of Bernard and Jones (1996a), who by means of a conclusion suggest "that international flows, associated mostly with Manufacturing, may not be contributing substantially to convergence either through capital accumulation or technological transfer" (Bernard and Jones 1996a: 1230). Our analysis suggests that this conclusion holds even stronger for manufacturing energy-productivity performance, where international flows cannot prevent an increase in cross-country differences of productivity levels.

The previous results raise the question as to what determines these cross-country productivity differences. In our search for an answer we subsequently take three steps. First, we go one step further in the σ-convergence analysis than Bernard and Jones (1996a, b) by examining productivity convergence for a breakdown of aggregate Manufacturing in order to see to whether the energy-productivity divergence and the lack of labour-productivity convergence observed in aggregate Manufacturing is also found within the different Manufacturing sub-sectors. Second, we perform a β-convergence analysis to test whether a statistically significant negative relationship exists between the initial level and the growth rate of productivity, in order to gain a better insight into the mechanism behind the observed convergence patterns. Third, we will try to explain differences in cross-country productivity growth by examining the role of different country-specific variables in driving energy- and labour-productivity growth at the sectoral level. The remaining part of this section is devoted to a σ-convergence analysis for a breakdown of aggregate Manufacturing into 10 sub-sectors. The other issues are the subject of Section 5.

The lower part of Table II presents the standard deviation of the log of, respectively, energy- and labour productivity for each of the 10 Manufacturing sub-sectors included in our dataset. The results reveal that the pattern of divergence in cross-country energy-productivity performance at the level of aggregate Manufacturing is to be found only in Iron and Steel and Non-Ferrous Metals. On the contrary, Food, Machinery, Non-metallic Minerals and Textiles all display evidence of (strong) σ-convergence. Cross-country productivity differences remain more or less constant in Chemicals, Transport Equipment, Paper and Wood. It can also be seen that the lack of labour-productivity convergence in aggregate Manufacturing is the result of mixed convergence patterns in different manufacturing sectors. Chemicals, Iron and Steel, Non-ferrous Metals and Wood exhibit (strong) convergence, while Machinery shows the opposite pattern of divergence. The sectors Food, Non-Metallic Minerals, Textile, Paper and Transport Equipment display no clear evidence for either convergence or divergence.

In conclusion, cross-country variation of energy-productivity is substantially higher than of labour-productivity at all levels of sectoral aggregation, and in particular in Services, Chemicals, Paper, Wood and at an ever increasing rate also in Iron and Steel and Non-Ferrous Metals. In Machinery, however, energy- and labour-productivity have strongly converged, resulting in a relatively small – although seemingly persistent – difference in the degree of cross-country variance. Moreover, convergence patterns turned out to depend on the level of aggregation, with different sectors displaying varying behaviour: some show reduction in variation, some increasing variation and others neither a clear reduction nor increase in cross-country differences.

These results suggest that different mechanisms may be at work in the different sectors. For example, the observed patterns of divergence might be the result of increasing international specialization while the tendency to converge might be caused by technology spill-overs from 'leaders' to 'followers', allowing lagging countries to catch up. Moreover, our results suggest that determinants of energy-productivity growth and labour-productivity growth might differ from each other, since we found no clear-cut (and sometimes even an opposite) relationship between cross-country convergence patterns in terms of energy productivity and labour productivity. Finally, even in those sectors showing evidence of convergence there remain substantial cross-country productivity differences, in particular in terms of energy productivity.

A possible explanation for the relatively high variation in energy-productivity levels across countries might be that cross-country differences in environmental awareness (influenced by social pressure) or stringency of environmental policies cause energy-efficiency improvements to be a matter of urgency at different degrees in different countries. Another reason might be a lack of international diffusion of energy-saving technologies as compared to technologies enhancing labour productivity. This can be caused by the fact that, in contrast with labour costs, in most sectors energy costs form only a small part of total production costs and, hence, firms do not have the incentive to search for best-practice technologies on the international market, as opposed to labour-augmenting technologies. Another explanation for the relatively high cross-country variation in energy-productivity levels might be the heterogeneity in energy mix across the OECD countries.

In any case, the observed cross-country variation in energy-productivity levels suggests that convergence does not pertain to a uniform steady state for all countries. In order to further examine this issue, we continue in the next section with a search for empirical regularities in the productivity improvements over our cross-section of countries by testing for sectoral patterns of β-convergence. As part of that analysis we will also try to explain (differences in) energy- and labour-productivity growth.

5. β-Convergence

The concept of β-convergence builds on the notion that countries that are further away from their steady-state level experience faster productivity growth. An empirical test thus builds on a regression of productivity growth on initial productivity. A negative correlation between the two provides an indication for convergence, because it suggests that countries with relatively low initial energy- and labour-productivity levels catch up to more advanced economies (see Section 2). A problem that one encounters in this respect is the quantitative characterization of the steady-state productivity level. Several approaches can be followed, each making different assumptions regarding the role of country-specific characteristics in driving productivity growth across countries. In this paper, we show the results for two types of analysis. First, we do a conditional convergence analysis, assuming productivity levels to converge towards multiple steady states that are conditional on (unspecified) country-specific characteristics.[10] Second, we try to identify the country-specific characteristics that determine (differences in) energy- and labour-productivity growth across countries.

Econometrically, we have estimated four different types of models, viz. a pooled Ordinary Least Squares model, a fixed-effects model, a random-effects model and a random-effects model with a Mundlak specification. On theoretical as well as on econometric grounds, there are good reasons to prefer the fixed-effects model. The OLS estimation method is valid only under the assumption that the error term is independent of the explanatory variables. However, in the growth regressions that we will estimate it is very likely that the error term contains all sorts of (unobserved) country-specific tangible and intangible factors that affect productivity growth.[11] As a result, OLS estimates tend to be biased and inconsistent in this case (Hsiao 1986). A panel approach applying fixed- or random-effects models can be used to solve this problem. This approach is capable of allowing for cross-country differences in steady states in the form of unobservable individual 'country-effects', thus diminishing the omitted-variables problem (Islam 1995). Comparing the fixed- and the random-effects model, the random-effects model uses up fewer degrees of freedom than the fixed-effects model and is conceptually appealing because of its characterization of the sources of the errors in a dataset with cross-section and time-series variation. However, in a growth context the requirement in a random-effects model of zero correlation between the individual country-effects and the observed explanatory variables is problematic, implying it to be an inadequate formulation in the context of our study. This problem can be solved by explicitly specifying the individual country-effects as a function of the variables with which it is supposedly correlated. This can be done by following the specification suggested by Mundlak (1978).[12]

In conclusion, there is reason to believe that the fixed-effects model or the random-effects model with Mundlak adjustment are to be preferred over the pooled OLS regression model and the normal random-effects model. For reasons of space constraints, in the remainder of this section we only report the results of the fixed-effects models, using the Least Squares Dummy Variables (LSDV) estimator. All other results for the four types of models that we have estimated – including specification tests that in almost all cases point at the fixed-effects model as the model to be preferred – can be found in the Annex to this paper. In Section 5.1 we report the results for the model in which the country-characteristics are not specified, viz. purely modelled as fixed effects. In Section 5.2 we go one step further and try to identify the country-specific characteristics that determine differences in energy- and labour-productivity growth across countries.

5.1. SECTORAL PATTERNS OF β-CONVERGENCE

As was just explained, we start our analysis of β-convergence by implementing a fixed-effects panel-data model for each sector, regressing the growth rate (g) of, respectively, energy- and labour productivity (y), on the log of its initial level (y_{t-1}) and unspecified country-specific (fixed) effects (α_i):

$$g_{it} = \alpha_i + \beta \ln(y)_{i,t-1} + \varepsilon_{it} \tag{4}$$

with i and t denoting, respectively, the cross-country and the time-series dimension. We assume ε_{it} to be an independently identically distributed random variable with mean 0 and variance σ_ε^2. Following Islam (1995) we use five-year time intervals in order to reduce the influence of business-cycle fluctuations and serial correlation of the error term. Hence, the growth rate (g) in equation (4) is an average over a five-year period. Because of notational ease we use the symbol y interchangeably for energy productivity (y_E) and labour productivity (y_L). The proper interpretation will be clear from the context.

In Tables IIIa and IIIb we present for each sector the estimated coefficient β obtained from equation (4) for energy- and labour-productivity, respectively, including various indicators and specification tests, which we will discuss below.[13] From Table IIIa it can be seen that we obtain a negative estimate of β for energy-productivity growth in all sectors, indicating the existence of β-convergence. Moreover, the estimate is statistically significant (at 1% significance level) in virtually all sectors.

Using the estimated values of β, the speed of convergence λ at which the productivity level is converging to a uniform productivity level can be calculated according to $\lambda = -[(1/T)\log(\beta+1)]$ with T denoting the length of the time interval under consideration, viz. 5 in this application. A convenient way of expressing this speed of convergence is the time t needed for the energy-productivity level to move halfway from its initial level (y_0) to the steady state productivity level y^*. This period of time is commonly referred to as the 'half life' (H).[14] The implied values of λ are also shown in Table IIIa.

It can be seen that the individual country effect explains between 16% (Machinery) and 98% (Wood) of the total unexplained variance, as indicated by ρ in Table IIIa. These results suggest that energy-productivity convergence depends to a large extent on individual country-effects, indicating convergence to be conditional rather than absolute in virtually all sectors. The estimated half life is between 1 year (Transport Equipment) and 14 years (Total). Of course, these

Table III. (a) β-convergence for energy productivity and (b) β-convergence for labour productivity

	Total	Manufacturing	Agriculture	Services	Transport	Chemicals	Food and Tobacco	Iron and Steel	Machinery	Transport Equipment	Non-Ferrous Metals	Non-Metallic Minerals	Paper	Textiles	Wood
a. β-convergence for energy productivity															
β	−0.221***	−0.616***	−0.480***	−0.218*	−0.630***	−0.262***	−0.518***	−0.389***	−0.231	−0.950***	−0.592***	−0.509***	−0.651***	−0.861***	−1.064***
	(0.07)	(0.07)	(0.09)	(0.12)	(0.16)	(0.08)	(0.13)	(0.11)	(0.14)	(0.11)	(0.14)	(0.10)	(0.10)	(0.13)	(0.19)
Implied λ	0.050	0.192	0.131	0.049	0.199	0.061	0.146	0.099	0.052	0.601	0.180	0.142	0.211	0.3949	NA
F-stat	10.26	82.22	29.15	3.48	15.65	9.83	16.07	12.21	2.85	71.07	17.26	26.96	39.72	44.95	30.03
Prob > F	0.00	0.00	0.00	0.07	0.00	0.00	0.00	0.00	0.10	0.00	0.00	0.00	0.00	0.00	0.00
R^2	0.41	0.76	0.61	0.44	0.42	0.33	0.36	0.32	0.31	0.68	0.38	0.55	0.60	0.60	0.60
ρ	0.65	0.88	0.45	0.28	0.88	0.51	0.78	0.56	0.16	0.77	0.73	0.42	0.93	0.62	0.98
# countries	13	14	14	9	12	13	13	13	13	13	12	13	13	13	12
# observ.	48	54	59	35	53	61	55	63	48	56	54	57	52	52	40
b. β-convergence for labour productivity															
β	−0.107***	−0.055	−0.083*	−0.178***	−0.104	−0.093*	−0.189***	−0.064	−0.054	−0.295**	−0.044	−0.209***	−0.105	−0.233***	−0.230***
	(0.03)	(0.04)	(0.04)	(0.04)	(0.11)	(0.05)	(0.05)	(0.11)	(0.05)	(0.12)	(0.09)	(0.06)	(0.08)	(0.05)	(0.07)
Implied λ	0.023	0.011	0.017	0.039	0.022	0.020	0.042	0.013	0.011	0.070	0.009	0.047	0.022	0.053	0.052
F-stat	16.66	2.09	3.71	17.86	0.87	3.68	12.03	0.37	1.23	5.98	0.22	10.83	1.81	18.28	12.52
Prob > F	0.00	0.15	0.06	0.00	0.36	0.06	0.00	0.55	0.27	0.02	0.64	0.00	0.19	0.00	0.00
R^2	0.68	0.33	0.39	0.80	0.16	0.29	0.49	0.11	0.45	0.22	0.10	0.39	0.24	0.33	0.47
ρ	0.48	0.28	0.31	0.62	0.07	0.19	0.44	0.06	0.37	0.21	0.08	0.28	0.17	0.20	0.22
# countries	14	14	14	8	11	13	13	13	12	10	13	12	13	13	13
# observ.	60	68	66	32	46	63	61	63	57	42	63	57	61	61	63

Standard errors in parentheses. Asterisks denote levels of significance: *** at 1%, ** at 5%, and * at 10%. ρ indicates the fraction of variance due to the fixed error component.

results raise the question as to what are the country-specific variables that apparently are so important in driving energy-productivity growth. Before returning to this question (in Section 5.2) we will look at the results of estimating equation (4) for labour-productivity. The results are presented in Table IIIb.

From Table IIIb it can be seen that also in terms of labour-productivity growth β is negative in all sectors. Moreover, these estimates are statistically significant in most sectors, with aggregate Manufacturing being an important exception. These results confirm the findings of Bernard and Jones (1996a) who also report strong evidence for convergence in Services, weak evidence in Agriculture and lack of labour-productivity convergence in Manufacturing. As compared to energy productivity, in most sectors the estimates of β are rather small, indicating that lagging countries catch up only slowly. The implied values for the speed of convergence (λ) confirm the finding of a slow rate of convergence: the time needed for labour productivity to move halfway from its initial level (y_0) to the steady state y^* varies from 47 years (Transport Equipment) to 77 years (Non-Ferrous Metals).

Similar to the results for energy productivity, the individual country effects explain a substantial part of the total unexplained variance, as indicated by ρ in Table IIIb. However, the evidence on conditional labour-productivity convergence is less clear-cut than it is for energy-productivity convergence. As compared to energy productivity (see Table IIIa), the individual country effects also play a smaller role in explaining total unexplained variance in all sectors, of course except for Services and Machinery. For labour productivity these percentages lie in between 6% (Iron and Steel) and 62% (Services). In conclusion, these results suggest labour productivity convergence to be also conditional rather than absolute in most sectors. The evidence on the role of country-specific characteristics is, however, more ambiguous than in the case of energy-productivity convergence. Apparently, in terms of labour productivity the variation in explanatory variables over time is relatively small as compared to cross-country differences.

As previously noted, β-convergence is a necessary but not a sufficient condition for σ-convergence. Our findings confirm that those sectors showing evidence of σ-convergence (see Section 4) also display evidence of β-convergence. However, the opposite is not necessarily true, as is illustrated for labour productivity by the sectors Machinery, Non-Metallic Minerals and Textiles: they pass the test for β-convergence without showing evidence of σ-convergence (see Table II). So, despite evidence of β-convergence, crosscountry differences in productivity levels remain to exist and even increase in some sectors. Clearly, country-specific variables do play an important role in explaining these patterns, as also shown by the presented evidence of β-convergence. Recall from Section 2 that several mechanisms may be at work, causing 'followers' to grow faster than 'leaders': advanced economies may suffer from diminishing returns, lagging countries may benefit from knowledge spill-overs, production processes may converge due to increasing competition, etcetera. On the other hand, persistent differences in, for example, energy prices or wages, investment shares or specialization patterns, may contribute to persistent or even increasing productivity differences across countries. In order to explain differences in cross-country energy- and labour-productivity growth, in the next section we extend our β-convergence analysis by including relevant country-specific variables that may explain cross-country variation in energy- and labour productivity growth.

5.2. Sectoral Determinants of β-Convergence

We search for country-specific sectoral determinants of energy- and labour-productivity growth by including a number of country-specific explanatory variables in the various regression models. We change the fixed-effects model in equation (4) as follows:

$$g_{it} = \alpha_i + \beta \ln(y)_{i,t-1} + \sum_{j=1}^{5} \gamma_j x_{it}^j + \varepsilon_{it} \tag{5}$$

with x_i^j the additional country-specific explanatory variables and all other variables defined as in equation (4). The specified explanatory variables x_i^j are defined at the sectoral level and include:

$$\text{Energy prices}: x_{it}^{1E} = \frac{(p_{E,t}+p_{E,t-1}+p_{E,t-2})}{3}$$

$$\text{Wages}: x_{it}^{1L} = \frac{w_t+w_{t-1}+w_{t-2}}{3}$$

$$\text{Investment share}: x_{it}^{2} = \frac{I}{Y}$$

$$\text{Openness}: x_{it}^{3} = \frac{XGS+MGS}{Y}$$

$$\text{Balassa index}: x_{it}^{4} = \frac{XGS_i / \sum_{i=1}^{14} XGS_i}{\sum_{s=1}^{10} XGS_{i,s} / \sum_{i=1}^{14}\sum_{s=1}^{10} XGS_{i,s}}$$

$$\text{Economies of scale}: x_{it}^{5} = \frac{Y_S}{\sum_{s=1}^{13} Y_S}$$

where sectoral indices are omitted for reasons of expositional clarity and with energy prices (x_{it}^{1E}) or wages (x_{it}^{1L}) included, respectively, in case of explaining energy-productivity growth or labour-productivity growth.

We expect energy prices and wages to be positively correlated with, respectively, energy- and labour-productivity growth. We took a 3-year moving average for the energy price and wages to avoid capturing the effect of short-term price fluctuations, assuming that investments in energy- and labour-augmenting technologies do respond to a structural trend in energy price/wage developments rather than to short term fluctuations. By including the investment share as an explanatory variable we test for the so-called embodiment hypothesis or a vintage effect, assuming that higher investment will contribute to increasing energy- and labour-productivity growth via technological change embodied in new capital goods (see, for example, Howarth et al. 1991; Mulder et al. 2003). We expect openness to have a positive impact on productivity growth, since an open sector faces relatively strong competition as well as exchange of knowledge, both of which we assume to have a stimulating effect on productivity growth. The Balassa index is an indictor measuring relative specialization patterns. We expect that if a country specializes in a particular sector, that that sector will be technologically relatively advanced, and hence we expect a positive effect on productivity. Finally, including an indicator for the relative size of a sector within a country captures the potential effect of economies of scale on productivity growth, assuming that a large sector is able to invest relatively much in R&D and in new capital goods and, hence, might be a technological leader displaying relatively high productivity growth rates.

In Table IV we present the results of regressing average energy-productivity growth rates on initial energy productivity levels and these additional explanatory variables, according to equation (5).[15]

From Table IV it can be seen that the estimates of β are again negative in all sectors, and that all these estimates are statistically significant as well (in most sectors at the 1% significance level). The speed of convergence measured by the half life lies now in between 1 year (Textiles) and 8 years (Machinery). Compared to the results presented in Table IIIa this means a higher speed of convergence in most sectors.

Concerning the additional explanatory variables, we find that the energy price has the expected (positive) sign in all sectors, while the positive impact of energy prices on energy-productivity growth is statistically significant in Chemicals, Iron and Steel, Non-Metallic Minerals and Paper. This result makes sense since these are energy-intensive sectors. The effect of the investment share, openness, specialization, and economies of scale on energy-productivity growth is, however, limited and with mixed positive and negative signs. Of these variables specialization, measured by the Balassa index, and economies of scale, measured by the relative size of a sector within a country, have the largest statistically significant effect on energy-productivity growth. The Balassa index has a statistically significant positive effect in Iron and Steel and Non-Metallic Minerals, and a statistically significant negative effect in Chemicals, and Paper. The economies of scale effects is statistically significant positive in Chemicals and Non-Metallic Minerals and statistically significant negative in Transport Equipment. We find the vintage effect to have a statistically significant positive effect in the Transport sector only. We use a Likelihood Ratio test to discriminate between the restricted model of equation (4) and the unrestricted model of equation (5), in order to verify whether the inclusion of the additional variables does make sense at all. The test results show that for the sectors Agriculture, Food, Machinery and Transport Equipment we cannot reject the hypothesis that the coefficients on the additional variables are jointly zero. Thus, in all other sectors the model has improved by including the additional explanatory variables, but indeed only to a limited extent. Together

Table IV. Determinants of β-convergence for energy productivity

	Agriculture	Services	Transport	Chemicals	Food and Tobacco	Iron and Steel	Machinery	Transport Equipment	Non-Ferrous Metals	Non-Metalic Minerals	Paper	Textiles	Wood
β	-0.808^{***}	-0.740^{***}	-0.390^{*}	-0.723^{***}	-0.776^{***}	-0.718^{***}	-0.342^{*}	-0.876^{**}	-0.382^{***}	-0.837^{***}	-0.518^{**}	-0.918^{***}	-1.457^{***}
	(0.14)	(−0.21)	(0.21)	(0.13)	(0.24)	(0.16)	(0.19)	(0.31)	(0.12)	(0.15)	(0.18)	(0.20)	(0.28)
Implied λ	0.330	0.269	0.099	0.257	0.299	0.253	0.0837	0.417	0.096	0.363	0.146	0.500	NA
P_E	0.095	0.910	0.054	1.235*	0.842	1.519	0.074	−0.485	0.540	−0.035	0.840*	0.742	−0.530
	(0.87)	(0.55)	(0.14)	(0.62)	(0.74)	(1.14)	(0.75)	(−0.55)	(0.76)	(1.50)	(0.43)	(0.87)	(0.42)
I/Y	−0.399	−0.9974	−0.133	0.008	−0.338	−0.583	2.806	0.789	−0.094	1.605	−0.015	0.619	-2.092^{**}
	(0.96)	(0.83)	(0.32)	(0.35)	(1.11)	(0.60)	(1.96)	(1.36)	(0.33)	(0.94)	(0.71)	(1.48)	(0.90)
Openness				0.010	0.038	0.001	0.009	−0.025	−0.020	−0.018	0.026	−0.030	0.035
				(0.03)	(0.06)	(0.02)	(0.04)	(0.05)	(0.03)	(0.11)	(0.06)	(0.03)	(0.04)
Balassa				-0.652^{*}	−0.334	0.305^{*}	0.794	0.967	0.011	−0.150	−0.025	0.160	0.021
				(0.32)	(0.31)	(0.15)	(0.64)	(0.60)	(0.06)	(0.17)	(0.08)	(0.18)	(0.05)
Y_i/Y	9.436	6.545^{*}	17.573^{***}	53.823^{***}	−3.206	−8.123	−1.379	−14.462	87.283	−26.782	14.965	−13.079	54.457
	(6.61)	(2.84)	(3.71)	(10.12)	(10.12)	(16.67)	(4.61)	(19.65)	(50.90)	(22.25)	(10.53)	(11.28)	(55.83)
F-stat	9.58	3.6	13.71	10.67	3	6.03	1.19	2.08	3.45	6.04	3.9	5.57	6.01
Prob > F	0.00	0.06	0.00	0.00	0.03	0.00	0.35	0.13	0.02	0.00	0.01	0.00	0.01
R^2	0.69	0.80	0.80	0.79	0.58	0.73	0.64	0.56	0.65	0.78	0.68	0.74	0.83
ρ	0.77	0.97	0.94	0.94	0.88	0.89	0.69	0.85	0.86	0.79	0.97	0.67	0.99
LR-test	3.38	9.11	28.14	35.56	8.46	9.39	7.42	7.1	10.98	11.2	11.45	5.15	14.41
Prob > χ^2	0.34	0.03	0.00	0.00	0.13	0.09	0.19	0.21	0.05	0.05	0.04	0.40	0.01
# countries	13	5	9	12	12	13	12	9	12	11	11	12	9
# observ.	46	17	31	38	37	42	38	28	39	35	33	36	24

Standard errors in parentheses. Asterisks denote levels of significance: *** at 1%, ** at 5%, and * at 10%. ρ indicates the fraction of variance due to the fixed error component.

with the fact that even after including additional explanatory variables, the individual country effect still explains between 66% (Textiles) and 97% (Services) of the total unexplained variance as shown by ρ, this suggests other country-specific factors than those currently included play an important role in driving cross-country energy-productivity growth patterns.

In Table V we present the results for labour-productivity growth.[16] The results reveal (again) negative estimates of β in all sectors. The obtained estimates are all statistically significant, except for Services and Transport Equipment. The speed of convergence measured by the half life lies now in between 1 year (Non-Metallic Minerals) and 99 years (Services). Compared to the results presented in Table IIIb this also means a higher speed of convergence in most sectors. We find that wages have the expected (positive) sign in all sectors except for Services, while the positive impact of wages on labour-productivity growth is statistically significant in all sectors except Services, Chemicals, and Non-Ferrous Metals. Like for energy productivity, the effect of investment share, openness, specialization, and economies of scale on labour-productivity growth is limited and with mixed positive and negative signs. Of these variables, economies of scale have the largest statistically significant effect on labour productivity growth, with statistically significant positive effects in Transport, Chemicals, Iron and Steel, Machinery, Paper and Wood, and a statistically significant negative effect in Services. We find the Balassa index to have a statistically significant positive effect in Non-Metallic Minerals, and a statistically significant negative effect in Food and Iron and Steel. The statistically significant effects of openness are positive in the Non-Metallic Minerals and Paper sector while negative in the sectors Chemicals and Wood. Finally, again the results do not give much support to the vintage effect, with Iron and Steel and Non-Ferrous Metals being the only sectors displaying a statistically significant positive effect, while the effect is negative in Agriculture, Food and Wood. Finally, also for labour productivity we find the individual country effect to explain a large fraction of the total unexplained variance, in spite of including a range of additional explanatory variables. However, contrary to energy-productivity growth, the results of the Likelihood Ratio test indicate that, except for Transport Equipment, we can reject the hypothesis that the coefficients on the additional variables are jointly zero. In other words, for labour-productivity, the regression model of equation (5) is a better approximation of our data than the restricted models of equation (4).

In conclusion, the extended β-convergence analysis presented in this section confirmed that energy- and labour-productivity convergence are conditional rather than absolute, but can only partly answer the question as to which are the country-specific determinants of productivity growth driving the observed convergence patterns. In short, higher energy prices and wages are found to stimulate, respectively, energy-productivity growth (in the energy-intensive sectors) and labour-productivity growth, while the role of specialization, economies of scale and particularly openness and investment share seems to be limited.

Table V. Determinants of β-convergence for labour productivity

	Agriculture	Services	Transport	Chemicals	Food and Tobacco	Iron and Steel	Machinery	Transport Equipment	Non-Ferrous Metals	Non-Metalic Minerals	Paper	Textiles	Wood
β	−0.281***	−0.034	−0.870***	−0.480***	−0.493***	−0.842*	−0.397***	−0.360	−0.802**	−0.913***	−0.581***	−1.042***	−0.656***
	(0.09)	(0.14)	(0.14)	(0.15)	(0.10)	(0.41)	(0.11)	(0.51)	(0.29)	(0.16)	(0.15)	(0.15)	(0.12)
Implied λ	0.066	0.007	0.408	0.131	0.136	0.369	0.101	0.089	0.324	0.488	0.174	NA	0.214
Wage	0.172***	−0.213	0.400***	0.043	0.286***	0.188	0.147***	0.182	0.169	0.363***	0.258***	0.686***	0.318***
	(0.06)	(0.12)	(0.07)	(0.07)	(0.05)	(0.12)	(0.04)	(0.23)	(0.09)	(0.07)	(0.06)	(0.11)	(0.06)
I/Y	−0.453	−0.057	−0.406	0.809	−0.443	1.490	0.072	0.094	−0.472	−0.407	−0.262	0.204	−0.777*
	(0.29)	(0.18)	(0.24)	(0.81)	(0.40)	(1.55)	(0.60)	(1.32)	(0.78)	(0.35)	(0.19)	(0.47)	(0.42)
Openness				0.004	−0.031	0.020	−0.009	0.027	0.026	0.143**	0.039	0.011	−0.010
				(0.03)	(0.02)	(0.05)	(0.02)	(0.05)	(0.03)	(0.06)	(0.03)	(0.01)	(0.03)
Balassa				−0.011	−0.175**	−0.178	−0.158	−0.924	0.035	0.316***	−0.031	0.005	−0.019
				(0.33)	(0.08)	(0.53)	(0.30)	(1.06)	(0.04)	(0.11)	(0.03)	(0.06)	(0.02)
Y_i/Y	0.164	0.700	10.971***	54.522**	0.922	−0.308	6.429**	10.085	100.344	−2.201	8.756**	−3.269	3.453
	(2.12)	(0.79)	(3.07)	(18.25)	(2.39)	(80.27)	(2.47)	(17.73)	(85.46)	(8.83)	(4.11)	(4.40)	(12.96)
F-stat	3.51	2.18	16.39	3.66	8.65	1.65	12.65	1.00	3.40	7.40	13.13	11.12	10.46
Prob > F	0.02	0.15	0.00	0.02	0.00	0.28	0.00	0.48	0.08	0.00	0.00	0.00	0.00
R^2	0.52	0.84	0.77	0.72	0.78	0.71	0.88	0.43	0.85	0.71	0.81	0.74	0.83
ρ	0.26	0.90	0.83	0.97	0.96	0.76	0.93	0.48	0.95	0.77	0.69	0.73	0.80
LR-test	15.40	5.49	46.77	24.19	41.25	9.21	52.78	9.67	16.22	32.25	60.05	45.07	38.87
Prob > χ^2	0.00	0.14	0.00	0.00	0.00	0.10	0.00	0.09	0.01	0.00	0.00	0.00	0.00
# countries	13	5	9	8	12	6	9	7	6	11	11	12	10
# observ.	56	19	37	27	48	18	35	23	18	44	44	48	39

Standard errors in parentheses. Asterisks denote levels of significance: *** at 1%, ** at 5%, and * at 10%. ρ indicates the fraction of variance due to the fixed error component.

6. Conclusions

This paper extends the existing empirical analyses of convergence patterns by providing a unique systematic comparison of energy- and labour-productivity convergence at a detailed sectoral level for 14 OECD countries, covering the period 1970–1997. A σ-convergence analysis revealed that the development of the cross-country variation in energy- and labour-productivity performance depends on the level of aggregation, with different patterns of productivity convergence and divergence across sectors. At the macroeconomic level we found evidence for energy-productivity divergence, driven by aggregate Manufacturing, as well as labour-productivity convergence, mainly driven by Services. The Manufacturing energy-productivity divergence turns out to be caused by the Iron and Steel and the Non-Ferrous Metals sectors. Moreover, despite a lack of evidence of labour-productivity convergence at the aggregate Manufacturing level, there is evidence of labour-productivity convergence in several Manufacturing sub-sectors, with Machinery as the most important exception in that it shows a clear pattern of divergence (in particular after 1985).

A β-convergence analysis, using a panel-data approach, led to the conclusion that in most sectors energy-productivity growth is relatively high in countries with relatively low initial productivity levels, while in several sectors this is also true for labour productivity. This result supports the hypothesis that relatively backward countries tend to catch up to more advances economies, in particular in terms of energy productivity, possibly because they can benefit from the experience and technologies developed by the countries operating at the forefront.

However, in spite of the evidence of convergence, cross-country differences in energy- and labour-productivity performance seem to be persistent. Our β-convergence analysis has shown convergence to be conditional on cross-country differences in steady-state characteristics. This is in line with the results of our σ-convergence analysis, which indicated that cross-country differences in productivity levels persist, even in those sectors that display a convergence pattern. Moreover, we found that the speed of energy-productivity convergence is in general higher than the speed of labour-productivity convergence. Nevertheless, at the same time cross-country differences in energy-productivity levels were found to be still substantially larger than cross-country differences in labour-productivity levels at all levels of sectoral aggregation.

In our search for the country- and sector-specific fundamentals determining these (differences in) energy- and labour-productivity developments, we found energy prices to stimulate energy-productivity growth in the energy-intensive sectors and we also found a positive relationship between wages and labour-productivity growth in most sectors. However, our data show the cross-country differences in wages to be considerably larger than cross-country differences in final energy prices (measured by the standard deviation of the log of each variable). Hence, they are not likely to explain the persistent relatively high cross-country differences in energy-productivity levels as compared to labour-productivity levels. In addition, we found specialization and economies of scale to contribute to energy- and labour-productivity growth in several sectors, while the investment share and openness play only a very limited role in explaining (cross-country) differences in energy- and labour-productivity growth. These results imply a need for additional research to further explain sectoral trends in energy-and labour-productivity growth across countries.

Acknowledgements

We gratefully acknowledge useful and stimulating comments on earlier versions of this paper by two anonymous referees of this journal as well as by Jeroen van den Bergh, Kornelis Blok, Lucas Bretschger, Reyer Gerlagh, Ton Manders, Hein Mannaerts, Machiel Mulder, Sjak Smulders, Paul Tang, Scott Taylor, Herman Vollebergh, and (other) participants of the Monte Verità Conference on Sustainable Resource Use and Economic Dynamics. We thank Mark Koetse, Gert-Jan Linders and Simonetta Longhi for help with the econometric software. Financial support of NWO as well as the hospitality of CPB Netherlands Bureau for Economic Policy Analysis is gratefully acknowledged.

Notes

1. The paper by Miketa and Mulder (M&M) follows the same approach as we develop in this paper, and builds upon the working paper version of this paper (Mulder and de Groot 2003b). The two papers differ in important respects. Apart from studying the manufacturing sector as is done in M&M, in this paper we also consider developments in agriculture, services, transport and a macroeconomic aggregate, allowing us to control for aggregation bias. Second, we simultaneously look at energy- and labour productivity. Third, whereas M&M include developing countries in their analysis, we focus on the OECD, allowing for a more detailed analysis with better-quality data. It enables us to show that even for a relatively homogenous group of countries, substantial cross-country differences exist. Finally, on a more technical note and in contrast to M&M, this paper employs sector-specific energy-price data, it considers more control variables in the conditional convergence analysis, and it employs PPP's instead of market exhange rates (as in M&M) to convert monetary variables into a common currency, which is clearly preferable for a convergence analysis.
2. Although Dollar and Wolff (1988, 1993) distinguish 28 sectors, they only present a labour-productivity convergence indicator for a few years and did not perform a regression analysis to test for convergence patterns.
3. An in-depth discussion of this literature is beyond the scope of this paper. For good surveys we refer to Abreu et al. (2005), Barro and Sala-i-Martin (1995), Broadberry (1996), Durlauf and Quah (1999), Fagerberg (1994), Economic Journal (1996) and Islam (2003).
4. Exploratory data analysis also does not provide much evidence for the existence of an 'Energy-intensity Kuznets Curve', with a period of increasing energy intensity preceding decreasing energy-intensity levels as countries get richer. See also Berndt (1978) for a review of long-term analysis of energy-productivity trends in the US, concluding that the (scarce) historical evidence of increasing energy intensity in the US in the period before 1910 is mainly attributable to limited data quality.

5. Based on information from the IEA Energy Balances (2001), one can analyze the cross-country dispersion of the share in total final energy consumption of the various energy types across the OECD countries (measured by the standard deviation of the log of these shares). Our own analysis reveals a clear trend towards decreasing cross-country variation in the use of the four main energy types (whereby natural gas and electricity are increasingly substituted for coal and oil). Hence, one might indeed expect this increasing homogeneity in energy mix to be an important source of cross-country energy productivity convergence within the OECD. Details of this analysis are available upon request.
6. Obviously, σ-convergence and β-convergence are closely related. A narrowing dispersion of cross-country productivity differences implies that countries with a relatively poor initial productivity performance tend to grow relatively fast. However, as has been argued by Quah (1993), a statistically significant inverse relationship between the initial *level* and the *growth rate* of productivity performance can be consistent with constant or even increasing cross-country productivity differences – a phenomenon known as Galton's Fallacy of regression towards the mean. We refer to Bernard and Durlauf (1996) and Durlauf and Quah (1999) for further discussion of empirical methodological issues of convergence tests.
7. For a detailed description of the dataset, we refer to an Annex to this paper that can be downloaded from http://www.henridegroot.net/pdf/annex_isdbe.pdf. The dataset can be downloaded as an EXCEL file from http://www.henridegroot.net/pdf/isdbe_dataset.xls.
8. In the literature on convergence analysis, two measures for σ-convergence are used interchangeably: (1) the standard deviation of the log of per capita income or productivity and (2) the coefficient of variation which equals the standard deviation of per capita income or productivity divided by the sample average. We have used both measures in our convergence analysis, finding both measures to yield an identical pattern of convergence, although with small differences in the size of cross-country variance. Details are available upon request. Here, we only present the result of the first measure.
9. Excluding Finland and Italy from the sample for Services reduces the cross-country dispersion by about 40% while leaving the pattern of σ-convergence unchanged.
10. We have also tested for unconditional convergence estimating a pooled Ordinary Least Squares Model, but all tests that we have performed clearly point at the relevance of conditional convergence. Details can be found in the Annex to this paper.
11. From the empirical macroeconomic growth literature – as briefly discussed in Section 2 – it is known that persistent differences in, for example, the technology level and institutions are an important factor in understanding cross-country differences in productivity and economic growth. Hence, any permanent unobserved factors would necessarily be correlated with the initial level of, respectively, energy- and labour productivity (y_{t-1}).
12. In his model, the individual country effect is assumed to be a linear function of the mean of the explanatory variables and a random country-specific effect, which is again assumed to be a random variable with mean zero and constant variance. As a result this formulation minimizes the bias induced by the correlation between individual effects and explanatory variables in a random-effects model – sometimes referred to as the heterogeneity bias (Chamberlain 1982). For space constraints, the results of this model are not reported in the main text. They can be found in the Annex to this paper.
13. The regression results as shown in Tables III and IV are based on an unbalanced panel, due to differences in data availability of the various variables per sector. For each sector we also list the number of observations and countries included in the regression. We have tested for the robustness of the presented β-convergence estimates by repeating the analysis for a balanced panel. This additional exercise showed that the exact results as reported in Tables III–V do change only slightly while the overall pattern of convergence and main conclusions still hold. Details are available upon request.
14. Approximating around the steady state, the convergence speed is given by $d\log(y_t)/dt = \lambda[\log(y^*) - \log(y_t)]$. Rewriting yields $\log(y_t) - \log(y_0) = (1 - e^{-\lambda t})[\log(y^*) - \log(y_0)]$

where (y_0) is the energy-or labour-productivity level at some initial date. From this equation we can derive that the half life (H) should satisfy the equality $e^{-\lambda H} = 0.5$, so $H = \ln(2)/\lambda$.

15. We also controlled for different specifications of energy prices (current prices, 5-year moving average, and log 3-year and log 5-year moving average), investment share ($(I/Y)_{t-1}$, (I/K), $(I/K)_{t-1}$ and $\ln(I/K)_{t-1}$), as well as an interaction term of investment share and log initial energy productivity ($\ln(Y/E)_0 * (I/Y)$). All these specifications did not substantially alter the estimates. Details are available upon request.
16. For labour productivity we also controlled for different specifications of the explanatory variables, including wages (current wage, 5-year moving average, and log 3-year and log 5-year moving average), investment share ($(I/Y)_{t-1}$, (I/K), $(I/K)_{t-1}$ and $\ln(I/K)_{t-1}$), as well as an interaction term of investment share and log initial labour productivity ($\ln(Y/E)_0 * (I/Y)$). All these specifications again did not substantially alter the estimates.

References

Abramovitz, M. (1986), 'Catching Up, Forging Ahead, and Falling Behind', *Journal of Economic History* **47**, 385–406.

Abreu, M., H. L. F. Groot and R. J. G. M. Florax (2005), 'A Meta-Analysis of β-Convergence: The Legendary 2%', *Journal of Economic Surveys* **19**(3), 389–420.

Barro, R. J. (1991), 'Economic Growth in a Cross-Section of Countries', *Quarterly Journal of Economics* **106**, 407–443.

Barro, R. J. and X. Sala-i-Martin (1992), 'Convergence', *Journal of Political Economy* **100**, 223–251.

Barro, R. J. and X. Sala-i-Martin (1995), *Economic Growth*, New York: McGraw Hill.

Bernard, A. B. and C. I. Jones (1996a), 'Comparing Apples to Oranges: Productivity Convergence and Measurement across Industries and Countries', *American Economic Review* **86**, 1216–1238.

Bernard, A. B. and C. I. Jones (1996b), 'Productivity across Industries and Countries: Time Series Theory and Evidence', *Review of Economics and Statistics* **78**, 135–146.

Bernard, A. B. and C. I. Jones (2001), 'Comparing Apples to Oranges: Productivity Convergence and Measurement across Industries and Countries: Reply', *American Economic Review* **91**, 1168–1169.

Bernard, A. B. and S. N. Durlauf (1996), 'Interpreting Tests of the Convergence Hypothesis', *Journal of Econometrics* **71**, 161–173.

Berndt, E. R. (1978), 'Aggregate Energy, Efficiency, and Productivity Measurement', *Annual Review of Energy* **3**, 225–273.

Broadberry, S. N. (1996), 'Convergence: What the Historical Record Shows', in B. van Ark and N. Crafts, eds., *Quantitative Aspects of Post-War European Growth.* Cambridge: Cambridge University Press.

Brock, W. A. and M. S. Taylor (2004), *The Green Solow Model*, Cambridge, MA: NBER Working Paper 10557.

Chamberlain, G. (1982), 'Multivariate Regression Models for Panel Data', *Journal of Econometrics* **18**, 5–46.

Chatterji, M. (1992), 'Convergence Clubs and Endogenous Growth', *Oxford Review of Economic Policy* **8**, 57–69.

Cleveland, C. J., R. K. Kaufmann and D. I. Stern (2000), 'Aggregation and the Role of Energy in the Economy', *Ecological Economics* **32**, 301–318.

Dollar, D. and E. N. Wolff (1988), 'Convergence of Industry Labour Productivity among Advanced Economies, 1963–1982', *Review of Economics and Statistics* **70**, 549–558.

Dollar, D. and E. N. Wolff (1993), *Competitiveness, Convergence, and International Specialization*, Cambridge, MA: MIT Press.

Durlauf, S. N. and P. A. Johnson (1992), *Local versus Global Convergence Across National Economies*, Cambridge, MA: NBER Working Paper 3996.

Durlauf, S. N. and D. Quah (1999), 'New Empirics of Economic Growth', in J. B. Taylor and M. Woodford, eds., *Handbook of Macroeconomics, Volume 1*. Amsterdam: North Holland Elsevier Science.

Economic Journal (1996), 'Controversy on the Convergence and Divergence of Growth Rates', *Special Issue Economic Journal* **106**, 1016–1069.

Fagerberg, J. (1994), 'Technology and International Differences in Growth Rates', *Journal of Economic Literature* **32**, 1147–1175.

Gerschenkron, A. (1952), 'Economic Backwardness in Historical Perspective', in B. Hoselitz, ed., *The Progress of Underdeveloped Areas*. Chicago: University of Chicago Press.

Grossman, G. M. and E. Helpman (1991), *Innovation and Growth in the Global Economy*, Cambridge, MA: MIT Press.

Howarth, R. B., L. Schipper, P. A. Duerr and S. Strøm (1991), 'Manufacturing Energy Use in 8 OECD Countries', *Energy Economics* **13**, 135–142.

Hsiao, C. (1986), *Analysis of Panel Data*, Cambridge: Cambridge University Press.

Islam, N. (1995), 'Growth Empirics: A Panel Data Approach', *Quarterly Journal of Economics* **4**, 1127–1170.

Islam, N. (2003), 'What Have We Learnt from the Convergence Debate?', *Journal of Economic Surveys* **17**, 309–362.

Jorgenson, D. W. (1984), 'The Role of Energy in Productivity Growth', *Energy Journal* **5**, 11–26.

Kaufmann, R. K. (2004), 'The Mechanisms for Autonomous Energy Efficiency Increases: A Cointegration Analysis of the US Energy/GDP Ratio', *Energy Journal* **25**, 63–86.

Keller, W. (2002), 'Geographic Localization of International Technology Diffusion', *American Economic Review* **92**, 120–142.

Lucas, R. E. (1988), 'On the Mechanics of Economic Development', *Journal of Monetary Economics* **22**, 3–42.

Miketa, A. (2001), 'Analysis of Energy Intensity Developments in Manufacturing Sectors in Industrialized and Developing Countries', *Energy Policy* **29**, 769–775.

Miketa, A. and P. Mulder (2005), 'Energy Productivity across Developed and Developing Countries in 10 Manufacturing Sectors: Patterns of Growth and Convergence', *Energy Economics* **27**, 429–453.

Mulder, P. and H. L. F. de Groot (2003a), '*International Comparisons of Sectoral Energy- and Labour-Productivity Performance. Stylised Facts and Decomposition of Trends*', CPB Discussion Paper 22, The Hague.

Mulder, P. and H. L. F. de Groot (2003b), '*Sectoral Energy- and Labour Productivity Convergence*', CPB Discussion Paper 23, The Hague.

Mulder, P., H. L. F. de Groot and M. W. Hofkes (2003), 'Explaining Slow Diffusion of Energy-Saving Technologies, A Vintage Model with Returns to Diversity and Learning-by-Using', *Resource and Energy Economics* **25**, 105–126.

Mundlak, Y. (1978), 'On the Pooling of Time Series and Cross Section Data', *Econometrica* **46**, 69–85.

Quah, D. (1993), 'Galton's Fallacy and Test of the Convergence Hypothesis', *Scandinavian Journal of Economics* **95**, 427–443.

Quah, D. (1997), 'Empirics for Growth and Distribution: Polarization, Stratification, and Convergence Clubs', *Journal of Economic Growth* **2**, 27–59.

Romer, P. M. (1986), 'Increasing Returns and Long-Run Growth', *Journal of Political Economy* **94**, 1002–1037.

Romer, P. M. (1990), 'Endogenous Technological Change', *Journal of Political Economy* **98**, S71–S102.

Schipper, L. and S. Meyers (1992), *Energy Efficiency and Human Activity: Past Trends and Future Prospects*, Cambridge: Cambridge University Press.

Solow, R. M. (1956), 'A Contribution to the Theory of Economic Growth', *Quarterly Journal of Economics* **70**, 65–94.

Sørensen, A. (2001), 'Comparing Apples to Oranges: Productivity Convergence and Measurement Across Industries and Countries: Comment', *American Economic Review* **91**, 1160–1167.

Swan, T. W. (1956), 'Economic Growth and Capital Accumulation', *Economic Record* **32**, 334–361.

Unander, F., S. Karbuz, L. Schipper, M. Khrushch and M. Ting (1999), 'Manufacturing Energy Use in OECD Countries: Decomposition of Long Term Trends', *Energy Policy* **27**, 769–778.

van Ark, B. (1993), *International Comparisons of Output and Productivity*, Monograph Series no.1, Groningen, Growth and Development Centre.

van Ark, B. and D. Pilat (1993), 'Productivity Level in Germany, Japan, and the United States: Differences and Causes', *Brookings Papers on Economic Activity. Microeconomics* **2**, 1–46.

9. Spatial Evolution of Social Norms in a Common-Pool Resource Game

JOËLLE NOAILLY[1,*], CEES A. WITHAGEN[2,3], AND JEROEN C. J. M. VAN DEN BERGH[2,4]

[1] *CPB Netherlands Bureau for Economic Policy Analysis, The Hague, The Netherlands;*
[*] *Author for correspondence (e-mail: noailly@cpb.nl; phone +31-70-338-3498; fax +31-70-338-3350)*
[2] *Department of Spatial Economics, Faculty of Economics and Business Administration, Free University, Amsterdam, The Netherlands;*
[3] *Faculty of Economics and Business Administration, Tilburg University, Tilburg, The Netherlands;*
[4] *Institute for Environmental Studies, Free University, Amsterdam, The Netherlands*

We study the conditions for the emergence of cooperation in a spatial common-pool resource (CPR) game. We consider three types of agents: cooperators, defectors and enforcers. The role of enforcers is to punish defectors for overharvesting the resource. Agents are located on a circle and they only observe the actions of their two nearest neighbors. Their payoffs are determined by both local and global interactions and they modify their actions by imitating the strategy in their neighborhood with the highest payoffs on average. Using theoretical and numerical analysis, we find a large diversity of equilibria to be the outcome of the game. In particular, we find conditions for the occurrence of equilibria in which the three strategies coexist. We also derive the stability of these equilibria. Finally, we show that introducing resource dynamics in the system favors the occurrence of cooperative equilibria.

Key words: common property, cooperation, evolutionary game theory, local and global interaction game, self-organization

1. Introduction

The common-pool resource (CPR) game is an excellent vehicle to study social dilemmas. A social dilemma is a situation in which the pursuit of individual interest comes at the expense of the collective goals. In the context of the management of common-pool resources, such a social dilemma results in overexploitation and inefficiency compared to the Pareto optimum.

Are people's actions always governed by selfish behavior? Recent evidence has led economists to reconsider their assumptions on behavior. In practice, a certain proportion of the population often exhibits cooperative behavior that seems in contradiction with a rational, selfish agent perspective. Such behavior is especially common when social norms prevail. These can operate in a decentralized way through a system of mutual trust, reward or punishment.

L. Bretschger and S. Smulders (eds.), Sustainable Resource Use and Economic Dynamics, 191–216.

Ostrom (1990) collected a large range of case studies of rural communities in which the presence of social norms led to sustainable management of common-pool resources. An example that has received much attention is the lobster industry in Maine. In this community, fishermen were assigned a spatial territory to spread their traps. In order to increase their catch, free-riders tried to expand their territory. Every fisherman, however, was allowed to defend his territory using different degrees of sanctions ranging from reprimands to opening or destroying the traps of the free-riders (Acheson, 1988). In other settings, ceasing cooperation with rather than punishment of free-riders has also proved effective. For example, Japanese villagers, Irish fishermen and inhabitants of the Solomon islands chose to cut contact with other members of the community who were overfishing (Taylor, 1987; McKean 1982; Hviding and Baines 1994). In this way, free-riders are deprived of the benefits provided by cooperation in other economic activities.

Next to case studies, there is also much experimental evidence that supports the persistence of cooperation. This literature is too large to be reviewed here. Seminal work has been done by Ostrom et al. (1994) and Fehr and Gächter (2001). The latter study shows that often a small proportion of 'altruistic punishers' in the population is sufficient to enforce cooperation in the group. Van Soest and Vyrastekova (2004) provide an application in the field of renewable resources.

A key theoretical question that follows from this is: Why does cooperative behavior emerge in the first place? Compared to the real world evidence there is not so much theory on this subject. Fehr and Schmidt (1999) develop a theoretical model of inequity aversion. They assume that a small proportion of people is willing to sacrifice material payoffs if this leads to more 'fair' and equitable outcomes. Sethi and Somanathan (1996) discuss the view expressed by Dasgupta (1993), who offers three possible explanations.

1. Small communities can be considered as mini states with the capacity to force members of the community to accept rules of behavior. Sethi and Somanathan (1996) do not find this a strong argument, because it cannot explain the fact that sanctioning by private individuals can be spontaneous and may entail destructive actions that are often prohibited at the state level.
2. Rationality in a repeated game can be reconciled with cooperation. This is the well-known Folk theorem. But the problem here is of course that the set of potential equilibrium outcomes is very large and that alternating periods of cooperation and defection can arise, contradicting observed persistence of strategies.
3. Social norms are internalized through "communal living, role modeling, education and through experimenting rewards and punishments" (Dasgupta, 1993, p. 208). They can then thus motivate agents to do what they do.

To address the problem associated with explanation 2 and analyze the solution offered under 3, adopting an evolutionary game setting is a promising option. By tracing the evolution of cooperation (and defection) it can help to determine which hypothetical equilibria with or without cooperation are actually feasible from a dynamic as well as from a disaggregate (population) perspective.

Theoretical models to explain or analyze the role of social norms to sustain cooperation in a resource setting are rare. Sethi and Somanathan (1996) aim to analyze which norms, as mentioned under point 3 above, can be internalized, using an evolutionary game theoretic framework. In their model, agents can choose between three strategies: defection, cooperation or enforcement. Agents who choose to be enforcer punish defectors, even though they incur a cost for doing so. The sanction level and the cost of sanctioning borne by defectors and enforcers depend on the number of defectors and enforcers in the population. Payoffs are related to the size of the resource stock and, for defectors (and enforcers), to the sanction (punishing) cost level. The agents can modify their strategy over time through a process of social learning. They learn by imitating the strategy that yields above average profits in the population. This is modeled by a replicator dynamics that mimics the evolution of social norms in the population. Sethi and Somanathan identify two main equilibria: a population composed of only defectors and a population composed of only cooperators and enforcers.

Another theoretical study of the role of social norms in solving social dilemmas is Eshel et al. (1998), who consider a model of local interactions between altruistic and egoistic agents. Although they do not deal with a resource, they nevertheless suggest relevant elements for our approach. In the first place, they assume that agents imitate the strategy in their direct neighborhood with the highest average profit. Second, they are able to derive analytical results for a setting in which agents are spatially distributed on a circle and interact only with their two nearest neighbors.

In the present paper, we consider a spatial evolutionary CPR game that combines both local and global interactions. Agents can be cooperators, defectors or enforcers, and imitate the strategy yielding above average payoffs in their neighborhood. We model space just like in Eshel et al. (1998) by assuming a circle with agents that only observe their two nearest neighbors. This is a logical conceptual-analytical starting point, while it also provides a quite accurate picture of how interactions occur in a large range of CPR issues, for example irrigation problems. Indeed, in many rural communities experiencing water conflicts, the monitoring of water quotas is exerted by the farmer located upstream or downstream of the water flow (see Ostrom 1990; Smith 2000), suggesting a linear (or circular to avoid edge problems) model. In line with this, we assume in our model that enforcers can only punish defectors located in their immediate neighborhood, which implies local interaction. Payoffs further depend on the aggregate harvesting effort and on the evolution of the stock of the resource, which means global interactions. In other words, our model combines local and global interactions. We derive theoretical and numerical results on type of limit states that emerge in such a system. We obtain two main innovative results compared to previous work. First, equilibria in which the three types of strategies coexist survive in the long-run. Second, the emergence of such equilibria, and of cooperative equilibria in general, is facilitated when resource dynamics is introduced.

The paper is organized as follows. Section 2 presents the standard CPR game and its evolutionary version. Section 3 sets out the main results obtained with our model for the case without resource dynamics. Section 4 discusses the stability of equilibria. Section 5 presents the results with resource dynamics. Section 6 concludes.

2. The CPR Game

We consider the performance of three types of agents: cooperators, defectors and enforcers. They play a game that involves the exploitation of a common pool of a renewable natural resource. Cooperators and enforcers are supposed to display social behavior, meaning that they restrict the level of harvesting effort exercised. Defectors, however, are only interested in their own profits, and harvest with a relatively high effort level, thereby possibly harming the other players. In order to be more precise with regard to these concepts we introduce here briefly the standard CPR game as a benchmark (see e.g., Dasgupta and Heal 1979; Chichilnisky, 1994; or Ostrom et al. 1994). We consider first the case of no resource dynamics. Subsequently we discuss the case where the natural resource changes over time. Then we introduce the evolutionary CPR game.

2.1. THE STANDARD CPR GAME

A fixed population of $n (n > 1)$ agents has access to a common pool of resources. Initially, we assume that the size of the pool is constant over time. The exploitation of the resource leads to harvest. The individual effort level of agent i is denoted by $x_i (i = 1, 2, \ldots, n)$. The individual cost of effort is denoted by w. Total effort is:

$$X = \sum_{i=1}^{n} x_i. \tag{1}$$

Harvest depends on individual as well as aggregate effort. When aggregate effort is X total harvest is equal to $F(X)$. It is assumed that F is strictly concave and increasing, $F(0) = 0$, $F'(0) > w$, and $F'(\infty) < w$. The harvested commodity is taken as the numeraire. Each agent i receives a share of total revenues equal to his share in aggregate effort. Individual profits are then given by:

$$\pi_i(x_i, X) = \frac{x_i}{X} F(X) - w x_i. \tag{2}$$

Aggregate profits are:

$$\Pi(X) = \sum_{i=1}^{n} \pi_i(x_i, X) = F(X) - wX. \tag{3}$$

The Pareto efficient, aggregate profit maximizing, level of effort is defined by $F'(X_{\mathrm{P}}) = w$. The zero profit level of efforts is defined by $F(X_0) = wX_0$. The symmetric Nash equilibrium aggregate effort follows from

$$\frac{(n-1)F(X_{\mathrm{C}})}{n \quad X_{\mathrm{C}}} + \frac{1}{n}F'(X_{\mathrm{C}}) = w. \tag{4}$$

Clearly $X_0 > X_{\mathrm{C}} > X_{\mathrm{P}}$. So, the Nash equilibrium is suboptimal, but yields positive rents.

In the case of resource dynamics the social optimum can be described in several ways. One option (in continuous time) is to consider the maximization of the present value of total profits

$$\max \int_0^\infty e^{-rt}[F(X(t), N(t)) - wX(t)]dt$$

subject to

$$\dot{N}(t) = G(N(t)) - F(X(t), N(t)), N(0) = N_0.$$

Here r is the social discount rate, $N(t)$ denotes the resource stock at time t, G is the natural growth function, and F is the harvest function, increasing in aggregate effort as well as in the existing stock. Social behavior can then be defined as behavior consistent with a dynamic extraction path that follows from present value maximization. The Nash equilibrium is the solution to the differential game where each agent takes the time path of efforts of all other players as given and maximizes his own total discounted profits.

2.2. THE EVOLUTIONARY CPR GAME

In the evolutionary CPR game a distinction is made between cooperators, defectors and enforcers. Defectors do not behave according to the social norm, and may be punished by enforcers. We first introduce the set of strategies. Next, we discuss the payoffs. Then, we go into the spatial structure of the game. Finally, we introduce replicator dynamics.

2.2.1. *Strategies*

In our evolutionary framework agents have a fixed strategy reflecting bounded rationality. The individual effort by cooperators and enforcers is denoted by x_{L} and the effort by individual defectors is x_{H}.

For the case of no resource dynamics it is assumed that these effort rates are constant and satisfy

$$X_{\mathrm{P}} \leqslant nx_{\mathrm{L}} < nx_{\mathrm{H}}. \tag{5}$$

Hence, if all players (n) are cooperators or enforcers they end up more closely to the Pareto efficient outcome than when all players are defectors.[1]

For the case of resource dynamics there are several plausible ways of modeling effort by individual agents. As suggested above, cooperation can be modeled by assuming that if all agents were cooperators, they would mimic the present value maximizing extraction path. A feature common to evolutionary approaches, however, is that agents use rules of thumb rather than adopt individually or socially optimal strategies. One way to capture this is to assume that effort rates of agents are constants, that may, however, differ across types of agents. For example, the individual effort of cooperators and enforcers is x_{L} with nx_{L} close to X^{pv}, defined as the steady state effort of the present value maximizing program, whereas effort by defectors is larger: $x_{\mathrm{H}} > x_{\mathrm{L}}$. If $nx_{\mathrm{L}} = X^{\mathrm{pv}}$ and all agents are cooperators, convergence to the present value optimal steady state ocurs. An alternative approach allows for the strategy to depend on the existing stock, in line with the work of Sethi and Somanathan (1996). They assume that all players can observe the existing resource stock, or are informed about the stock by an agency. Then one can define $x_{\mathrm{L}}(t) = \alpha_{\mathrm{L}} N(t)$ and $x_{\mathrm{H}}(t) = \alpha_{\mathrm{H}} N(t)$ with α_{L} and α_{H} positive constants with $\alpha_{\mathrm{H}} > \alpha_{\mathrm{L}}$. In particular, α_{L} can be chosen such that convergence occurs to N^{pv}, the present value maximizing steady state resource stock. It need not be the case that the socially optimal steady state coincides with the steady state arising from present value maximization. Other objectives than present value maximization can be pursued as well.

2.2.2. *Payoffs*

The numbers of cooperators, defectors and enforcers are denoted by n^{C}, n^{D} and n^{E}, respectively. All cooperators and enforcers exercise an effort level of $x_{\mathrm{L}}(N)$ (obviously the argument N can be suppressed when resource dynamics is not taken into account) each. Enforcers punish defectors, at a cost γ per detected defector. Defectors make an effort $x_{\mathrm{H}}(N)$ and pay a sanction δ per enforcer that detects them. Define $Z(X, N) = F(X, N)/X - w$, which can be interpreted as aggregate profit per unit of effort. Individual profits, can be written as follows:

$$\pi^{\mathrm{C}}(X, N) = x_{\mathrm{L}}(N) Z(X, N), \tag{6}$$

$$\pi_k^{\mathrm{D}}(X, N) = x_{\mathrm{H}}(N) Z(X, N) - \delta k, \tag{7}$$

$$\pi_m^{\mathrm{E}}(X, N) = x_{\mathrm{L}}(N) Z(X, N) - \gamma m. \tag{8}$$

Here $\pi_k^{\mathrm{D}}(X, N)$ denotes the profits of a defector punished k times and $\pi_m^{\mathrm{E}}(X, N)$ is the payoff of an enforcer punishing m times.

2.2.3. *Spatial structure*

Sethi and Somanathan (1996) assume that all enforcers in the population can detect all defectors and punish them. Formally, this means that $k = n^{\mathrm{E}}$ and $m = n^{\mathrm{D}}$. Obviously, the spatial structure is irrelevant then. In contrast, we assume that an enforcer can only detect and punish a defector in his immediate neighborhood. This calls for a definition of neighborhood. There are several straightforward ways to do so. Eshel et al. (1998) describe players as located on a circle, implying that every agent has exactly two direct neighbors. Hence k and m take the values 0,1, or 2. One could extend the notion of neighborhood to two positions on the circle at each side. Then k and m run from 0 to 4. Another convenient way of defining neighborhood is on a torus. A torus is a two dimensional lattice whose corners are pasted together to ensure that all cells are connected, so that there are no edge effects. Then an agent's neighbors are, for example, those to the west, east, north and south. In this case k and m run from 0 to 4. One could include also those to the north–east etc., at the cost of higher complexity. In the present paper we focus on the circle with each agent having two neighbors, because this allows us to derive interesting theoretical results that are much more difficult to obtain for the torus. For an extensive numerical analysis on the two-dimensional torus, using a different learning rule as well, we refer to Noailly et al. (2004).

The sanctioning cost falling upon an enforcer is proportional to the number of defectors detected and punished, which expresses the efforts made by the enforcer. Similarly, in our setup it matters by how many enforcers a defector is detected. In the case of two enforcers, the cost to the defector is twice as high as in the case of only one enforcer. This can be regarded either as reflecting the sum of the damages inflicted upon the defector by individual enforcers or as the level of punishment being dependent on the amount of evidence provided by all enforcers together.

2.2.4. *Replicator dynamics*

A common element of evolutionary game theory is replicator dynamics, describing when, how and why agents switch strategies. In Sethi and Somanathan (1996) agents are assumed to be able to observe their own profits and the average profits in the population. The decision to change strategy is based on the comparison of these profits. This gives rise to a replicator dynamics equation of the following form:

$$\dot{n}^j = n^j(\pi^j - \overline{\pi}),\ j = \mathrm{C, D, E} \tag{9}$$

where $\overline{\pi} = (n^{\mathrm{C}}\pi^{\mathrm{C}} + n^{\mathrm{D}}\pi^{\mathrm{D}} + n^{\mathrm{E}}\pi^{\mathrm{E}})/n$, the average payoff in the entire population at time t. Therefore, agents do not necessarily switch to the most profitable strategy instantaneously. It follows that an equilibrium with all three strategies, a so-called CDE-equilibrium (with Cooperators, Defectors and Enforcers) will never prevail, because in such an equilibrium enforcers would do strictly worse than cooperators. In contrast to Sethi and Somanathan we explicitly take into account that agents do not observe the payoffs of the entire population. We make the more realistic assumption that agents only observe the payoffs of all agents in their neighborhood, including themselves. The aggregate replicator dynamics formulation then has to be dropped. Several alternative imitation or selection mechanisms can be adopted. One is that an agent imitates the strategy in his neighborhood with the highest payoff. The advantage of this rule is its simplicity. But it can lead to outcomes that may be considered implausible. Consider, for example, the case where a cooperator is surrounded by two defectors, one not being punished (and better off than the cooperator) and the other one severely punished, paying a very high sanction. In such a case it might not be considered very plausible for the cooperator to switch to defection. On the torus, with a cooperator surrounded by three defectors, one of which is not punished and the other three severely punished, the example might even be more appealing. However, there are no fundamental objections against modeling the imitation dynamics in this way. An alternative approach is to switch to the strategy that is doing best on average in the neighborhood. This implies a certain degree of rationality on behalf of the agent. Applying this rule to the previous example, the cooperator becomes a defector if on average the defectors in the cooperator's neighborhood do better than the cooperator. This is the rule employed by Eshel et al. (1998) and we will use it the present paper too.

3. No Resource Dynamics

This section deals with the case where resource dynamics is not taken into account. Consequently, the variable N, denoting the resource stock, is suppressed. At any instant of time τ the system is characterized by the number of agents of each type, $n^{\mathrm{C}}(\tau)$, $n^{\mathrm{D}}(\tau)$ and $n^{\mathrm{E}}(\tau)$, summing up to the given number n, and by the location of each agent on the circle. For convenience, we fix one position on the circle and call it position 1. Then a state of the system can be represented by a vector of length n consisting of ordered C's, D's and E's. So, with $n = 5$, the notation CDEDE means that there is a cooperating agent at position 1, there are defectors at positions 2 and 4, and enforcers at positions 3 and 5 (note, however, that this state is essentially the same as DEDEC). Time is considered discrete. At time $\tau+1$ the system finds itself in a new state, as a consequence of agents switching from one strategy to another. In first instance strategy changes occur only on the basis of replicator dynamics. Mutation is studied in Section 4. The questions we address in the present concern the limiting behavior of the system, as τ goes to infinity.

We have been able to identify a rich set of limit states. First of all there are equilibria. A state is called an equilibrium if no agents wants to change strategy. Second, there are blinkers. A state is called a blinker if agents change strategy, but the new resulting state is a rotation of the original state. For example: the state characterized by CDEED is a blinker, if, after all agents have made their choice of strategy, the new state is DCDEE. So, essentially neither the numbers of cooperators, defectors and enforcers, nor their relative positions on the circle have changed. We also found cycling, where composition of the population of strategies as well as locations change over time, but where after one period the system reproduces.

As shown by the profit equations given in the previous section, payoffs are affected by both local and global factors, namely sanctioning among neighbors and aggregate efforts, respectively. The combination of these two types of factors is an innovative feature of the present paper. However, it entails the inconvenience to render the model much more complex to analyze. Under some assumptions with regard to the ranking of profits, general theoretical results can be derived for equilibria and blinking. With regard to cycling we restrict ourselves to providing an example to show that it can actually occur.

3.1. EQUILIBRIA AND BLINKERS

We aim to derive conditions for the existence of certain types of equilibria and blinkers. Profit rankings are not unambiguous: we might have $\pi_1^{\mathrm{E}}(X) < \pi_1^{\mathrm{D}}(X)$ for some values of X and $\pi_1^{\mathrm{E}}(X) > \pi_1^{\mathrm{D}}(X)$ for other values. This complicates a theoretical analysis and makes it difficult to obtain clear-cut results. Therefore, we concentrate on unambiguous profit rankings here. To avoid clutter we omit the argument X when there is no danger of confusion. For example, $\pi_0^{\mathrm{D}} > \pi^{\mathrm{C}}$ means $\pi_0^{\mathrm{D}}(X) > \pi^{\mathrm{C}}(X)$ for all relevant X (i.e., $nx_{\mathrm{L}} \leqslant X \leqslant nx_{\mathrm{H}}$). To allow for a theoretical approach, we assume that the following three sets of profit rankings hold:

1.
$$\pi_0^{\mathrm{D}} > \pi^{\mathrm{C}} = \pi_0^{\mathrm{E}}.$$
$$\pi_0^{\mathrm{E}} > \pi_1^{\mathrm{E}} > \pi_2^{\mathrm{E}}.$$
$$\pi_0^{\mathrm{D}} > \pi_1^{\mathrm{D}} > \pi_2^{\mathrm{D}}.$$
These rankings derive from the fact that we neglect the case of negative profits. This rules out the possibility that defectors do worse than cooperators even if they are not punished. Profits from harvesting are nonnegative if $Z(nx_{\mathrm{H}}) \geqslant 0$, because Z is decreasing and $X \leqslant nx_{\mathrm{H}}$.

2.
$$\pi^{\mathrm{C}} > \pi_1^{\mathrm{D}}.$$
In order to have an interesting game, being punished should not be uniformly more profitable than cooperation. Several choices are open regarding the number of punishments needed to make cooperation more profitable than defection. For simplicity, we assume that being punished once is already worse than being cooperative.

3.
$$\pi_1^{\mathrm{D}} > \pi_1^{\mathrm{E}} \text{ implies } \pi_1^{\mathrm{E}} > \pi_2^{\mathrm{D}} > \pi_2^{\mathrm{E}}.$$
$$\pi_1^{\mathrm{E}} > \pi_1^{\mathrm{D}} \text{ implies } \pi_1^{\mathrm{D}} > \pi_2^{\mathrm{E}} > \pi_2^{\mathrm{D}}.$$
Therefore, if being punished once is better than punishing once, then being punished twice is worse than punishing twice, and vice versa. Hence, in the former case, being a defector is not too advantageous.

We get analytical results for the set of parameter values that satisfy these assumptions, but the simulations suggest that the results we obtain analytically also hold for a much broader class of parameter values.

Since the imitation rule that we employ is based on comparison of average payoffs by agents, an additional distinction can be made. A defector punished once is doing better than an enforcer punishing once, with a non-punishing enforcer in his neighborhood, or this ranking is the other way around. To illustrate the intuition, consider the following complete string EEEDD, where the second defector is next to the first enforcer. The first and the third enforcers, both located next to a defector that is punished once, change to defection when the sanction rate is sufficiently low. However, with what we will call a moderately low sanction rate they stay enforcers.

From the above discussion and assumptions, we get the following profits orderings as stated in Definition 1.

Definition 9.1

(i) The sanction rate is relatively low if:
$\pi_0^D > \pi^C = \pi_0^E > \pi_1^D > \pi_1^E > \pi_2^D > \pi_2^E$.

(ii) The sanction rate is relatively very low if:
$\pi_0^D > \pi^C = \pi_0^E > \pi_1^D > \pi_1^E > \pi_2^D > \pi_2^E$ and $\pi_1^D > \frac{1}{2}\left(\pi_0^E + \pi_1^E\right)$.

(iii) The sanction rate is relatively moderately low if:
$\pi_0^D > \pi^C = \pi_0^E > \pi_1^D > \pi_1^E > \pi_2^D > \pi_2^E$ and $\pi_1^D < \frac{1}{2}\left(\pi_0^E + \pi_1^E\right)$.

(iv) The sanction rate is relatively high if:
$\pi_0^D > \pi^C = \pi_0^E > \pi_1^E > \pi_1^D > \pi_2^E > \pi_2^D$. So, the sanction rate is relatively low if $\pi_k^D > \pi_k^E$ for $k = 1, 2$. It is relatively high if $\pi_k^D < \pi_k^E$ for $k = 1, 2$. It should be noted that the wording, including 'relatively,' is chosen on purpose. For example, the sanction rate could be called absolutely low if $\pi_2^D > \pi_1^E$, or even $\pi_2^D > \pi_0^E$. We will consider such cases later on in this paper when performing simulations. Below we derive a set of sufficient conditions for each of the two rankings to hold, thereby showing that the definitions are not void.

Lemma 1

(i) *Suppose* $\gamma > \delta$ *and* $(x_H - x_L)Z(nx_L) < 2\delta - \gamma$. *Then the sanction rate is relatively low.*

(ii) *Suppose* $\gamma > \delta$, *and* $\delta - \frac{1}{2}\gamma < (x_H - x_L)Z(nx_H) < (x_H - x_L)Z(nx_L) < 2\delta - \gamma$. *Then the sanction rate is relatively very low.*

(iii) *Suppose* $\gamma > \delta$ *and* $(x_H - x_L)Z(nx_L) < \delta - \frac{1}{2}\gamma$. *Then the sanction rate is relatively moderately low.*

(iv) *Suppose* $(x_H - x_L)Z(nx_L) < \delta - \gamma$ *and* $(x_H - x_L)Z(nx_H) > \delta - 2\gamma$. *Then the sanction rate is relatively high.*

Proof. The proof of the lemma is given in the appendix.

The proof of the lemma is rather technical, but the idea behind it is easily explained. Consider, for example, statement (i). If the cost of sanctioning γ is higher than the sanction δ, then a defector being punished k times is better off than an enforcer punishing k times for all k, because profits from harvesting are higher for a defector, and the defector incurs a lower sanction than the cost the enforcer has to make to punish. Moreover, if $(x_H - x_L)Z(nx_L) < 2\delta - \gamma$, then $x_H Z(X) - 2\delta < x_L Z(X) - \gamma < 0$ for all $X \leqslant nx_H$ and hence $\pi_1^E > \pi_2^D$. All the other proofs follow the same approach.

A further distinction suggests itself: a relatively very high versus a moderately high sanction rate, according to $\frac{1}{2}\left(\pi_0^{\mathrm{D}}+\pi_1^{\mathrm{D}}\right)$ being smaller or larger than π_1^{E}, respectively. However, this distinction is not meaningful, as can be seen as follows. The inequality $\frac{1}{2}\left(\pi_0^{\mathrm{D}}+\pi_1^{\mathrm{D}}\right) < \pi_1^{\mathrm{E}}$ requires $(x_{\mathrm{H}}-x_{\mathrm{L}})Z(X) < \frac{1}{2}\delta-\gamma$ for all $X \leqslant nx_{\mathrm{H}}$, so that it is necessary that $\frac{1}{2}\delta-\gamma > 0$. But the inequality $\pi_1^{\mathrm{D}} > \pi_2^{\mathrm{E}}$ requires $(x_{\mathrm{H}}-x_{\mathrm{L}})Z(X) > \delta-2\gamma = 2(\frac{1}{2}\delta-\gamma)$. This is a contradiction. Also, note that the relatively high sanction rate implicitly assumes that $\delta > \gamma$, since $(x_{\mathrm{H}}-x_{\mathrm{L}})Z(nx_{\mathrm{L}}) > 0$.

Next we establish several propositions regarding the existence and the characteristics of equilibria and blinkers, assuming that the profit ranking satisfies one of the definitions given above. States with only cooperators ('allC'), only defectors ('allD'), only enforcers ('allE'), and only cooperators and enforcers ('CE'), are always an equilibrium. A state with only defectors and cooperators ('CD') cannot be an equilibrium, because a cooperator next to a defector will change to defection. Therefore, we concentrate on the DE and CDE equilibria. A cluster in an equilibrium is a string of adjacent agents playing identical strategies. To start with we prove a lemma that turns out to be rather helpful.

Lemma 2 *Suppose $n \geqslant 3$.*

(i) *A string composed as CED cannot occur in an equilibrium.*
(ii) *A string composed as CD cannot occur in an equilibrium.*
(iii) *A string composed as DED cannot occur in an equilibrium.*
(iv) *A string composed as EDE cannot occur in an equilibrium.*

Proof.

(i) With CED, the punishing enforcer switches to cooperation, if not to defection.
(ii) With CD the defector switches to cooperation or the other way around.
(iii) and (iv) Obviously, DED cannot occur under a relatively low sanction rate, and EDE is ruled out in the case of a relatively high sanction rate. If DED would occur in an equilibrium with a relatively high sanction rate, the defectors surrounding the enforcer would not be punished twice, since EDE is ruled out. But then the enforcer would switch to defection. To exclude EDE in the relatively low sanction case, the same type of argument holds.

Proposition 1. *Suppose the sanction rate is relatively very low.*

(i) *There exists neither a DE nor a CDE equilibrium.*
(ii) *There exists neither a DE nor a CDE blinker.*

Proof.

(i) Suppose there exists an equilibrium with $n^{\mathrm{E}} > 0$ and $n^{\mathrm{D}} > 0$. There must be at least one enforcer next to a defector, because the equilibrium does not consist of defectors only, and if a defector is not punished, he cannot be a neighbor of a cooperator, because then the cooperator switches to defection. If a defector next to an enforcer is punished only once the enforcer will switch to defection, because $\pi_1^D > \frac{1}{2}(\pi_0^E + \pi_1^E)$, a contradiction. Hence every defector is punished twice, contradicting lemma 2(iv).

(ii) Suppose there is a blinker with $n^E > 0$ and $n^D > 0$. At least one agent switches to enforcement. This is not a cooperator. So, a defector should switch to enforcement. He will only do so if he is punished twice: so we have EDE. In order for the first enforcer in this string to switch to defection, we need DEDE, because with EEDE he will stay an enforcer. But now the first defector in the row will never switch to enforcement. This proves statement (ii) of proposition 1.

Proposition 2. *Suppose the sanction rate is relatively moderately low.*

(i) *For a DE-equilibrium to obtain it is necessary that $n \geqslant 5$. If $n = 5$ the equilibrium configuration is given by EEEDD. In any DE-equilibrium enforcers occur in clusters of minimal length 3.*

(ii) *For a CDE-equilibrium to obtain it is necessary that $n \geqslant 9$. If $n = 9$ the equilibrium configuration is given by CEEEDDEEE. In any CDE equilibrium any enforcer adjacent to a defector is part of a cluster of at least 3 enforcers.*

(iii) *There exists neither a DE nor a CDE blinker.*

Proof. The proof of the proposition is given in the appendix.

The intuition behind the proposition is straightforward. Since, by definition, $\pi_1^E < \pi_1^D < \frac{1}{2}\left(\pi_0^E + \pi_1^E\right)$, punishing enforcers need to be 'protected' by non-punishing enforcers. This leads to clusters of three enforcers. Protection by cooperators does not work, because, in an equilibrium, a punishing enforcer can never be located next to a cooperator. This also explains why a minimal number of players is required. Obviously, it might be the case that in a CDE-equilibrium the majority of agents is defecting.

Proposition 3. *Suppose the sanction rate is relatively high.*

(i) *For a DE-equilibrium to obtain it is necessary that $n \geqslant 5$. If $n = 5$ the equilibrium configuration is given by EEDDD. In any DE-equilibrium defectors occur in clusters of minimal length 3.*

(ii) *For a CDE-equilibrium to obtain it is necessary that $n \geqslant 8$. If $n = 8$, the equilibrium configuration is given by CEEDDDEE. In any CDE-equilibrium any defector adjacent to an enforcer is part of a cluster of at least 3 defectors.*

(iii) *There exist no DE blinkers. There do exist CDE blinkers. A necessary condition is $n \geqslant 4$. If $n = 4$, the blinker is CDDE.*

Proof.

(i) and

(ii) The proof of statements (i) and (ii) follows the lines of the proof of the previous proposition. It will not be given here.

(iii) Non-existence of DE blinkers is obvious. Suppose $n = 3$ and there is a CDE blinker. Then the cooperator remains a cooperator. Both the enforcer and the defector turn into cooperators. Hence there is no blinking in this case. Suppose $n = 4$. In a CDE blinker a cooperator never becomes an enforcer. Hence, at least one cooperator should turn into a defector. This can only be the case if he is next to a defector who is not punished. In the present case we cannot have CDCE because both cooperators will become defectors. Hence the only equilibrium candidate is CDDE. It is easily verified that this is a blinking equilibrium.

At this stage, we can summarize the main existence properties of the equilibria. We have established that the states C, D, E and CE are always part of equilibria, while CD never is. We also have proved that DE and CDE equilibria only occur for a moderately low sanction rate and for a sufficiently large population. Finally, we have shown that DE blinkers only occur for a high sanction rate and a sufficiently large population.

3.2. CYCLING

To illustrate the phenomenon of cycling in the present setting, consider the following initial state: DDDDEE. The defectors in positions 2 and 3 will not change strategy. The first and fourth defector change strategy if the average payoff of the defectors in their neighborhood is smaller than the payoff of an enforcer punishing once:

$$\frac{1}{2}\left[\pi_0^{\mathrm{D}}(X) + \pi_1^{\mathrm{D}}(X)\right] < \pi_1^{\mathrm{E}}(X). \tag{10}$$

If this inequality holds, for $X = 2x_{\mathrm{L}} + 4x_{\mathrm{H}}$, the enforcers stick to enforcement since then also

$$\pi_1^{\mathrm{D}}(X) < \pi_1^{\mathrm{E}}(X). \tag{11}$$

Therefore, if (10) holds, the new state becomes EDDEEE. The enforcers at positions 1 and 6 in the new state switch to defection if

$$\frac{1}{2}\left[\pi_0^{\mathrm{E}}(X) + \pi_1^{\mathrm{E}}(X)\right] < \pi_1^{\mathrm{D}}(X) \tag{12}$$

for $X = 4x_{\mathrm{L}} + 2x_{\mathrm{H}}$. When this condition holds, the defectors stay defectors. Now set $x_{\mathrm{L}} = 100$, $x_{\mathrm{H}} = 120$, $F(X) = 13.25X^{1/2}$, $w = 0.5$, $\gamma = 0.1$, $\delta = 0.525$. Then all conditions are satisfied. Therefore cycling between the two states indicated above, occurs with a period of one. It may be noticed that the range of the sanction δ, given the other parameter values, is rather small. This small range is also found in various other numerical examples with different parameter values for x_{L}, x_{H} and the parameters of F. It suggests that cycling does not occur for a wide range of parameter values. Obviously, this does not matter, since the aim was just to provide an example. Moreover, it would be relatively easy to induce cycling if we allow profits from harvesting to be negative: $Z(X) < 0$. In this case the incentive of defectors to change strategy is much larger for defectors, because they earn less from harvesting than enforcers (they incur greater losses). In our example we took care that profits, even including sanctions and the cost of sanctioning, are positive. The importance of the example is that it shows that the system is not only steered through local interaction, but that global interaction through aggregate efforts plays a role too.

Comparing the results in this section with those obtained by Sethi and Somanathan, we observe that we not only have more types of limit states (cycling, blinking and equilibria), but within the class of equilibria, we have equilibria with cooperation surviving next to defection, which is a novel finding as well. This phenomenon occurs for sanction levels that can be deemed realistic. So, it turns out that the spatial structure of the game is pivotal in the characterization of potential equilibria.

4. Stability

In the previous section we have established the existence of equilibria where cooperators survive in groups with many defectors. This result is due to the spatial structure of our model. It would be less interesting if the occurrence of these equilibria would merely be a coincidence, namely for very specific spatial constellations, or if the equilibria would easily be disrupted by players making mistakes in choosing their strategies. In the present section we investigate this issue. We first make use of an approach common in applications of evolutionary game theory. Then we discuss and explore an alternative route, relying on numerical simulations with stochastic features.

In evolutionary game theory stability of equilibria is tied to mutations, meaning that players may make mistakes in deciding on their strategy. This then leads to the notion of stochastic stability. Before dealing with stochastic stability in detail we illustrate the concept by means of an example. Suppose we start with a configuration of only cooperators. This configuration will persist if all players strictly follow the imitation rule. However, suppose that each player has a given small probability of making a mistake. At some instant of time this probability materializes and a player becomes a defector. Then defection will infect a large part of the population within finite time: many cooperators will be eradicated. And it is highly unlikely that the stochastic process of mutation will restore the 'allC' equilibrium. This is essentially why this equilibrium is not stochastically stable.

One way to assess the stochastic stability or instability of equilibria is outlined in Young (1998) and in Eshel et al. (1998). We briefly sketch the procedure, merely to illustrate the difficulties encountered in its application. As was stated before, at any instant of time τ the state of the system is characterized by the number of agents of each type, $n^{\mathrm{C}}(\tau)$, $n^{\mathrm{D}}(\tau)$ and $n^{\mathrm{E}}(\tau)$, summing up to the given number of agents n, and by the location of each agent on the circle. Such a representation may be misleading, however. If two states are identical up to rotation or taking the mirror image, they should be considered as identical states. For example: the state CCDDEEE is essentially the same as CDDEEEC (each player is moved one position) and as EEEDDCC (we 'read' the circle in the opposite direction). So, in the sequel, we restrict ourselves to unique states. The state space is the finite set of all possible states. The matrix P of transition probabilities p_{ij} from state i to state j, is completely determined by the imitation dynamics. To keep things simple, we assume that a situation where a player has two equivalent strategies to choose from does not occur. Then the transition matrix consists of zeros and ones only. Next, we introduce mutation. After the transition to a new state a player has a probability $\frac{1}{2}\alpha$ of not adopting the strategy that is optimal according to the imitation rule, but, instead, going to pursue either of the two alternative strategies. So, a player who just became a cooperator, according to the imitation rule, will actually act as a defector or an enforcer, each with probability $\frac{1}{2}\alpha$. This yields another matrix of probabilities denoted by Q with a typical element q_{ij} denoting the probability of transition from state i to state j, as a consequence of the mutations that happen to take place in state i. The overall transition matrix is then γ with $\gamma_{ij} = \sum_k p_{ik} q_{kj}$. Let μ be the solution of the following system $\mu\,\gamma = \mu$, where μ is on the unit simplex: $\mu \geqslant 0$ and $\sum_i \mu_i = 1$. The vector μ is the unique stationary distribution of the process for a given mutation rate. Element μ_i indicates that as time gets large, state i will occur during a proportion μ_i of time. Finally, one considers the limit of μ for the mutation rate approaching zero.

It is clear from the exposition given above that in the case at hand it is almost unsurmountable to derive general results on the stochastic stability of CDE equilibria in our model. Already for the minimal number of agents in the low sanction case the set of possible states amounts to hundreds. Eshel et al. (1998) were able to derive results on stochastic stability, thanks to the fact that their analysis only involves two strategies. Moreover, Sethi and Somanathan (1996) do not inquire into stochastic stability, arguing that: "Given the time scales relevant for this paper, the introduction of stochastic perturbations is therefore unlikely to affect our main inferences." Like in the case of Sethi and Somanathan, one might consider our model as applying to fisheries. The time scales can be interpreted as referring to seasons, while updating occurs once per season. If an equilibrium would not persist after, say, 1000 seasons, then this should not be considered as a sign of instability because it concerns an extremely long time horizon for the system considered. In other words, if it takes thousands of seasons and thus years before a certain type of equilibrium (e.g., CDE) has completely vanished, then from a practical perspective this should not be regarded as a serious case of instability. Indeed, many other, directed factors will then have ample time to exercise their influence on the system and its stability, negating the relevance of the stochastic factors.

In view of the previous argument we investigate stability of the different equilibria, and in particular of CDE-equilibria, using numerical simulations. We employ the harvest function given by $F(N, X) = N^{1/2}X^{1/2}$ and consider a population of $n = 100$ agents. The other parameter values are

$$w = 5, \quad N_0 = 10^6, \tag{13}$$

$$x_H = 120, \quad x_L = 100, \tag{14}$$
$$\delta = 280, \quad \gamma = 300. \tag{15}$$

These parameters are chosen such that $nx_L = X_P$, implying that when all agents harvest low the social optimum is reached. Further, we have $Z(nx_H) > 0$, so that in the absence of sanctioning all players enjoy positive profits. In a first step, we illustrate the above statement of Sethi and Somanathan (1996) by studying the time scales on which cooperative equilibria cease to occur. We start from a fixed spatial configuration, namely a CDE initial state with $n^C = 25, n^D = 25$ and $n^E = 50$. The agents are positioned in the following order: 25 cooperators, 25 enforcers, 25 defectors and 25 enforcers. In the absence of mutation and with $\delta = 280$, this initial state is a CDE equilibrium. How does the frequency of CDE equilibria evolve when we introduce mutations? We assume that in every round each agent has a probability of making a mistake of $\alpha = 5/1000$, meaning that, at the beginning of every round, the agent has a chance of α to deviate from the decision rule. We record the population configuration at the end of every round. We conduct 100 simulation runs for different time horizons and compute the average time spent in each possible population configuration. The results are reported in Table I.

After 10,000 rounds, the system spent on average 24% of the time in a CDE-configuration. As expected, as the time horizon increases, i.e., as the number of mutations rises, the frequency of CDE-equilibria decreases. Eventually, as $\tau \to \infty$, the frequency will tend to zero. Nevertheless, this frequency decreases by only 1% per additional 10,000 rounds. After 30,000 rounds, the system spends still 22% of the time in a CDE-equilibrium. This suggests that the time scales over which CDE disappears may be very long and irrelevant for applications with seasonal updating. Note also that the mutation rate is kept constant in this experiment, whereas it should converge to zero in a proper test for stochastic stability.

Our approach with spatial interaction lends itself to examine stability of equilibria in an alternative manner, namely to look at the emergence of equilibria and the frequency of the different types of equilibria when we randomize over the initial shares of strategies as well as their distribution over the circle. For a given sanction rate δ, we vary:

1. the initial shares of each strategy in the population. To reduce the number of runs necessary to cover all the possible combinations of initial shares, only

Table I. Percentage of time spent in each equilibrium in the presence of mutations

τ	D-equil.	DE-equil.	CE-equil.	CDE-equil.
100	0.00	0.01	0.00	0.99
500	0.13	0.27	0.00	0.60
10,000	0.58	0.18	0.00	0.24
20,000	0.59	0.18	0.00	0.23
30,000	0.56	0.22	0.00	0.22

strategy shares that are multiples of 0.05 are considered. The set of initial coordinates $Z = ((1; 0; 0), (0.95; 0.05; 0) \ldots (0; 0; 1))$ is composed of coordinates $z_0 = (n^C/n, n^D/n, n^E/n)$. Further, we eliminate initial strategy shares composed of only cooperators and defectors, and of only cooperators and enforcers, as the outcomes can be easily predicted in these cases.[2] This leaves us with 190 potential initial shares,

2. the initial spatial distribution of strategies. For every z_0, we perform 100 so-called runs of 200 time-steps.[3] Each run starts with a draw from a uniform random spatial distribution, such that the probability of a position on the circle being occupied by a player of type j equals $n^j/n (j = \mathrm{C, D, E})$. This means that for each z_0, we consider 100 random spatial arrangements and register the resulting equilibrium. We find that on average 32% of the runs (out of 19,000) converge to a D-equilibrium, 4% converge to a CE-equilibrium, 33% to a DE-equibrium and 29% to a CDE-equilibrium. Cycling occurred in the CDE-configuration in 2% of the cases. We found no occurrence of blinker states. This is in line with our theoretical results since the sanction level $\delta = 280$ corresponds to a relatively moderately low sanction rate. What can we conclude from the fact that in almost 30% of the cases convergence to a CDE-equilibrium occurs? Formally, it does not prove the stochastic stability of this type of equilibrium. But the procedure followed strongly suggests that CDE-equilibria are not a mere coincidence. In an environment that is stochastic with respect to initial shares and initial locations, cooperation will survive in a large number of cases.

Additionally, these simulations provide two other types of insights on how the system works. First, we gain insights on how the initial distribution affects equilibria. Figure 1 shows the frequency of convergence to each equilibrium for the different initial shares combinations. In each graph, each z_0 is represented by a dot. The grey-black scale indicates the result of simulations with 100 random spatial distributions after 200 time steps. A black colored coordinate indicates that, starting with the respective z_0, all runs converge to the given type of equilibrium.[4] As expected, D-equilibria are more easily achieved for initial populations with few enforcers and, inversely, CE-equilibria are more likely to be reached for initial populations composed of many enforcers. CDE-equilibria are most frequently achieved for middle-range initial shares with a slight majority of enforcers.

Second, we gain insights on the effects of the initial location of strategies over space. Figure 2 shows the evolution of strategy shares over time starting from three identical share vectors $z_0 = (0.30; 0.30; 0.40)$ but with different initial spatial arrangements. The evolution of strategy shares is governed by two forces. First, enforcers who punish a lot imitate defectors in their neighborhood. In some sense, enforcers are then being eliminated by defectors. Second, enforcers who punish at least one defector switch to cooperation when cooperators are located in their neighborhood. So, we see that enforcers have a hard life. On the other hand, they

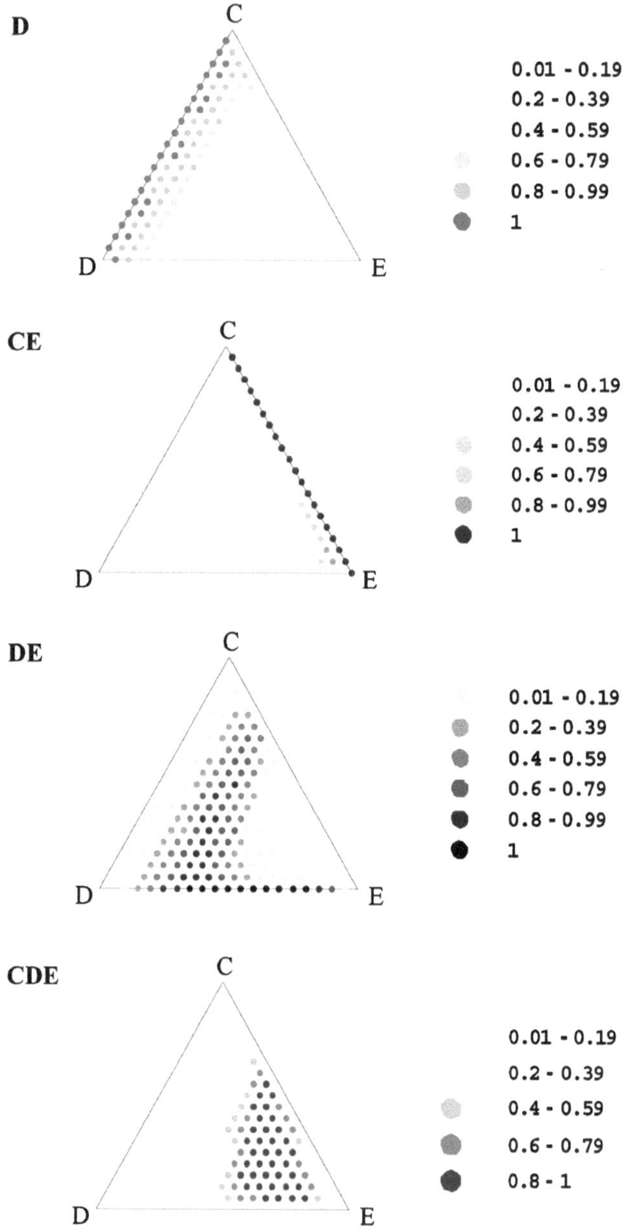

Figure 1. Frequency of equilibria for different initial shares multiple of 0.05, $\delta = 280$.

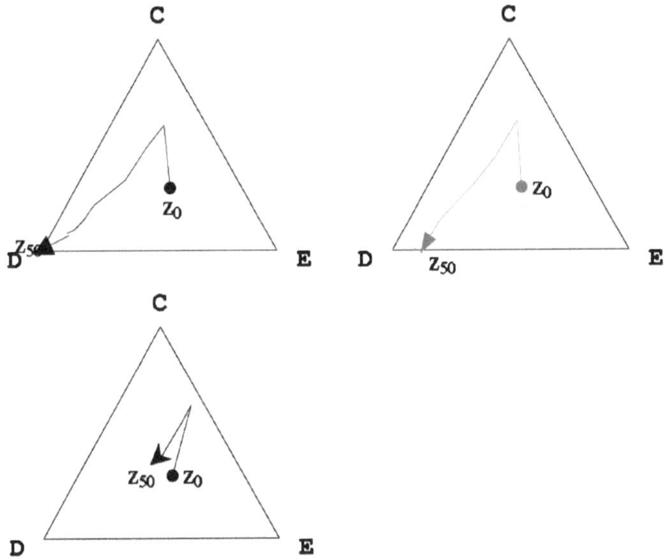

Figure 2. Evolution of strategy shares over time, for $z_0 = (0.30; 0.30; 0.40)$, $\delta = 280$.

eliminate defectors if they punish hard enough. In all of the approach paths we see the number of enforcers decrease; the number of defectors increases in the final steps.

Finally, to complete our analysis of stability and to confirm further that the occurrence of CDE-equilibria is not a mere coincidence, we run simulations for various sanction levels. Given our parameter values, the definition of a relatively very low sanction is satisfied for $200 < \delta < 232$. The sanction rate is relatively moderately low if $232 < \delta < 341$. It is relatively high if $400 < \delta < 680$. We also performed simulations for sanction rates outside the ranges that imply an unambiguous ordering of profits. For each sanction level, we performed 19,000 simulation runs and computed the average frequence of occurrence of each equilibrium. The results are displayed in Figure 3. The exact frequencies for each type of equilibrium can be found in Table B.1 in Appendix B.

As expected, the frequency of D-equilibria decreases as the sanction rises. Inversely, the frequency of CE-equilibria increases with the sanction level. The largest frequency of CDE-equilibria is found for $\delta = 700$. Beyond $\delta = 800$, the frequency of CE-equilibria rises sharply and it becomes almost impossible for defectors to survive in the population, as shown by the fall in the frequency of CDE- and D-equilibria. As expected from proposition 3, we also find blinkers in the range of relatively high sanction rates, even if the occurrence of this phenomenon is relatively rare (see Table B.1 in Appendix). Recall that for a CDE blinker to occur, the sanction level should be high and a single enforcer should be located between a cooperator and a defector. In large populations this is unlikely to happen. We also find that the occurrence of cycling CDE-equilibria is quite rare. The main conclusion we can draw from these exercises is that equilibria with cooperation have a high probability of survival.

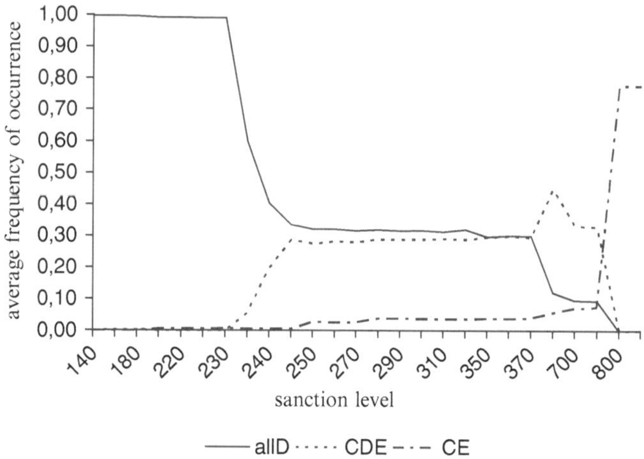

Figure 3. Average frequency of D-, CE-, and CDE-equilibria for different sanction levels.

5. Resource Dynamics

The role of resource dynamics on harvesting behavior is often neglected in the literature on common-pool issues. Experiments and games developed by Ostrom et al. (1994) do not pay any attention to resource dynamics. In real-world situations, however, harvesters are likely to reconsider and actually modify their strategies on the basis of observed changes in the resource stock. Feedback effects are present from harvesting activities to the natural resource and vice versa. Resource dynamics raises the issue of the dynamic development of the resource itself and the impact of varying resource stock level on harvest. In addition, a new dynamic issue is relevant in the present context, namely how resource dynamics affects the occurrence of cooperation. We start the analysis by postulating a logistic natural growth function:

$$G(N) = \rho N\left(1 - \frac{N}{K}\right) \tag{16}$$

with ρ the intrinsic growth rate and K the carrying capacity. Harvest is oftentimes represented by the Shaefer function where the harvest rate is effort multiplied by the resource stock. Alternatively, we assume that

$$F(X, N) = X^{\beta} N^{1-\beta} \tag{17}$$

with $0 < \beta < 1$. Updating of the resource stock after each round follows the usual pattern:

$$N_{t+1} = N_t + G(N_t) - F(X_t, N_t). \tag{18}$$

The steady state of the system is then the solution of

$$\rho N\left(1 - \frac{N}{K}\right) = X^{\beta} N^{1-\beta}. \tag{19}$$

We follow Sethi and Somanathan (1996) and assume that individual effort is proportional to the existing resource stock in the following manner:

$$x_{\mathrm{H}} = a_{\mathrm{H}} N \tag{20}$$

$$x_{\mathrm{L}} = a_{\mathrm{L}} N. \tag{21}$$

What is the effect of the introduction of resource dynamics on the limit states? It is to be expected that the qualitative nature of the limit states will not change: blinkers, cycling and equilibria can still occur. In Section 3.2 we saw that cycling resulted from the fact that payoffs are affected by aggregate harvest. Similarly, resource dynamics will influence payoffs, increasing the number of situations under which profit reversal and thus cycling will occur. In other words, given that an additional global interaction mechanism is operative, cycling is likely to become more frequent. We further expect that the likelihood of the occurrence of CDE-equilibria will not decrease. Overharvesting as a consequence of higher effort levels by defectors does not only reduce harvesting profits per unit of effort but also through the resulting smaller resource stock itself. Therefore, with a given effort rate of defectors, being a defector becomes relatively less rewarding when there are many defectors.

In the case of resource dynamics we can write:

$$\pi^{\mathrm{C}} = a_{\mathrm{L}} N \left[\left(\frac{1}{n^{\mathrm{D}}(a_{\mathrm{H}} - a_{\mathrm{L}}) + n a_{\mathrm{L}}} \right)^{\beta} - w \right], \tag{22}$$

$$\pi_k^{\mathrm{D}} = a_{\mathrm{H}} N \left[\left(\frac{1}{n^{\mathrm{D}}(a_{\mathrm{H}} - a_{\mathrm{L}}) + n a_{\mathrm{L}}} \right)^{\beta} - w \right] - k\delta, \tag{23}$$

$$\pi_m^{\mathrm{E}} = a_{\mathrm{L}} N \left[\left(\frac{1}{n^{\mathrm{D}}(a_{\mathrm{H}} - a_{\mathrm{L}}) + n a_{\mathrm{L}}} \right)^{\beta} - w \right] - m\gamma. \tag{24}$$

Consider π_k^{D}. We see that if n^{D} increases, two things happen. First, aggregate profits from harvesting given by the term in square brackets in (22) decrease. This is similar to the no resource dynamics case: it is a consequence of higher efforts, given the stock. Second, the stock decreases (after some time). This also leads to smaller profits as an additional effect. The stock effect can be assessed by realizing that the steady state with n^{D} enforcers equals:

$$N(n^{\mathrm{D}}) = K \left(1 - \frac{(n^{\mathrm{D}}(a_{\mathrm{H}} - a_{\mathrm{L}}) + n a_{\mathrm{L}})^{\beta}}{\rho} \right). \tag{25}$$

So, the stock effect comes in addition to the effort effect.

We run simulations with a_{H} and a_{L} fixed so that we can compare the average frequency of occurrence of equilibria with the case without resource dynamics. We fix $a_{\mathrm{L}} = 0.0001$ and take $\delta = 300$, $K = 2^{*}10^{6}$ and $\rho = 0.2$. For the rest we employ the same parameters as before. This yields a steady state stock of 10^6 if all players were cooperators or enforcers. The parameter value $a_{\mathrm{L}} = 0.0001$ corresponds with $x_{\mathrm{L}} = 100$ while $a_{\mathrm{H}} = 0.0002$ corresponds with $x_{\mathrm{H}} = 200$ in the

Table II. Average frequency of convergence with and without resource dynamics, $\delta = 300$

With resource dynamics $x_{\mathrm{H}} = a_{\mathrm{H}}N$	D-equil.	DE-equil.	CDE-equil.
200	1.00	0.00	0.00
300	0.77	0.19	0.03
350	0.60	0.34	0.07
400	0.51	0.40	0.09
No resource dynamics			
200	1.00	0.00	0.00
300	1.00	0.00	0.00
350	0.65	0.32	0.03
400	0.55	0.39	0.07

case without resource dynamics. We calculate the frequency of equilibria for these parameter values with resource dynamics as well as without resource dynamics. In both cases D-equilibria occur with probability one. Similarly we performed the simulations for higher values of a_{H}. The results are given in Table II. We find that for identical x_{H}, resource dynamics leads to increasing occurrence of CDE-equilibria, as expected.[5]

Finally, we can show that in the case of fixed effort rates, the same type of results is to be expected. With fixed effort rates x_{L} and x_{H} we get

$$\pi^{\mathrm{C}} = x_{\mathrm{L}} \left[\left(\frac{N}{n^{\mathrm{D}}(x_{\mathrm{H}} - x_{\mathrm{L}}) + n x_{\mathrm{L}}} \right)^{\beta} - w \right]$$

$$\pi_k^{\mathrm{D}} = x_{\mathrm{H}} \left[\left(\frac{N}{n^{\mathrm{D}}(x_{\mathrm{H}} - x_{\mathrm{L}}) + n x_{\mathrm{L}}} \right)^{\beta} - w \right] - k\delta$$

$$\pi_m^{\mathrm{E}} = x_{\mathrm{L}} \left[\left(\frac{N}{n^{\mathrm{D}}(x_{\mathrm{H}} - x_{\mathrm{L}}) + n x_{\mathrm{L}}} \right)^{\beta} - w \right] - m\gamma$$

Now the steady state stock is a bit less straightforward to calculate. It satisfies

$$\rho N \left(1 - \frac{N}{K} \right) = N^{\beta} (n^{\mathrm{D}}(x - \mathrm{H} - x_{\mathrm{L}}) + n x_{\mathrm{L}})^{\beta}.$$

It is not clear beforehand that this N is increasing in n^{D}. In fact it is increasing if and only if $\frac{N}{K} < \frac{1}{3}$. For this reason the case at hand is slightly more complicated. But, under this condition, essentially we see the same mechanism at work. Higher n^{D} decreases aggregate profits directly through the effort effect, and, in addition, decreases aggregate profits through its effect on the stock. All this implies that the difference $\pi_k^{\mathrm{D}} - \pi_m^{\mathrm{E}}$ decreases when n^{D} increases, and more than in the absence of resource dynamics.

6. Conclusions

This paper has studied the emergence of cooperation in a particular spatial CPR game, namely with space modeled as a circle. The combination of evolution, space and resource dynamics can lead to a complex model system that easily defies analytical solutions. Here we proposed a model that allowed derivation of various analytical results, while additional conjectures were supported by a large number of numerical simulations.

The major contribution of the present paper is that in the CPR game a cooperative strategy can survive, even when the majority of agents is defecting. This result runs counter to Sethi and Somanathan (1996). Our finding is due to the assumption that agents base their actions on the observed profitability of strategies employed by neigboring agents. In such a setting cooperators and enforcers can in some sense protect each other. By means of several types of simulations we were able to establish support for the view that cooperative equilibria are likely to persist, even in stochastically changing environments. Introducing resource dynamics reinforces our results.

From a conceptual perspective, the approach adopted here can be understood as combining local and global interactions. Virtually all related, analytical work in the literature has focused solely on local interactions, which evidently renders much simpler models. The global interactions in this case are due to two factors. First, profits are affected by aggregate harvest, to which all agents contribute. Second, profits depend on the resource stock, which changes due to the composition of harvesting strategies in the population of agents. The presence of global feedback means that profit rankings of strategies are not necessarily fixed over time. Indeed, due to changes in the composition of the population of strategies the aggregate harvest and resource stock change, which in turn may alter the conditions under which the agents interact. The important implication is that resource dynamics combined with spatial evolution increases the frequency of stable equilibria in which resource use is sustainable.

The analytical results apply mainly to the case without global interactions. The alternative case was illustrated by a combination of analytical results, illustrative examples and systematic numerical simulations. Evidently, future work might concentrate on extending the boundary of analytical findings.

Future research may be devoted to examining alternative redistribution schemes of the fines collected, at least if this is the interpretation given to the sanctions rather than damages incurred. It has been assumed thusfar that redistribution in lump sum. An alternative assumption would be that enforcers get some kind of compensation. Another item worth investigating in more detail is the distinction between a cooperator and a non-punishing enforcer. In the present approach the distinction cannot be made on the basis of actual behavior or payoffs. But for the analysis it does make a difference whether an agent is a cooperator or an enforcer. Therefore, this line of research would investigate the issue of signaling characteristics.

Acknowledgments

The authors would like to thank two anonymous referees, Charles Figuières, and the participants of several conferences and seminars: SURED (Ascona 2004), Workshop on Spatial Dynamic-Models of Economic and Ecosystems (Triest 2005), BIOECON (Cambridge 2005), UFSIA (Antwerp 2005) and UQAM (Montreal 2005).

Notes

1. One could be more specific by assuming for example that $X_P = nx_L < X_C \leqslant nx_H \leqslant X_0$.
2. When there are no enforcers in the population, defectors always earn more than cooperators and will spread quickly through the population. When there are no defectors in the population, cooperators and enforcers earn the same payoffs and stick to their strategies so that there is no further evolution of strategies.
3. Convergence to equilibria always occurred within 200 time steps.
4. For illustration purposes, we add the frequencies in all the extreme cases in which the initial population is composed of two strategies only.
5. With the given parameters, CE-equilibria do not occur.

References

Acheson, J. (1988), *The Lobster Gangs of Maine*, Hanover: University Press of New England.

Chichilnisky, G. (1994), 'North-South Trade and the Global Environment', *American Economic Review* **84**(4), 851–874.

Dasgupta, P. (1993), *An Inquiry into the Well-Being and Destitution*, Oxford: Oxford University Press.

Dasgupta, P. and G. Heal (1979), *Economic Theory and Exhaustible Resources*, Cambridge: Cambridge University Press.

Eshel, I., L. Samuelson and A. Shaked (1998), 'Altruists, Egoists and Hooligans in a Local Interaction Model', *American Economic Review* **88**, 157–179.

Fehr, E. and S. Gächter (2002), 'Altruistic Punishment in Humans', *Nature* **415**, 137–140.

Fehr, E. and K. Schmidt (1999), 'A Theory of Fairness, Competition, and Cooperation', *Quarterly Journal of Economics* **114**, 817–868.

Hviding, E. and G. Baines (1994), 'Community-Based Fisheries Management, Tradition and Challenges of Development in Marovo', *Solomon Islands Development and Change* **25**, 13–39.

McKean, M. (1992), 'Management of Traditional Common Lands (Iriaichi) in Japan', in D. Bromley and D. Feeny, eds., *Making the Commons Work* (pp. 66–98). San Francisco: ICS Press.

Noailly, J., Van den Bergh, J. and Withagen, C. (2004), Local and Global Interactions in an Evolutionary CPR Game on a Torus, Department of Spatial Economics, Free University Amsterdam.

Ostrom, E. (1990), *Governing the Commons: The Evolution of Institutions for Collective Action*, Cambridge: Cambridge University Press.

Ostrom, E., R. Gardner and J. Walker (1994), *Rules, Games and Common-Pool Resources*, Ann Arbor: University of Michigan Press.

Sethi, R. and E. Somanathan (1996), 'The Evolution of Social Norms in Common Property Resource Use', *American Economic Review* **86**(4), 766–789.

Smith, R. (2000), Institutional innovations among the Mormons: Collective Action in Irrigation. Working paper, Department of Political Science, Indiana University.

van Soest, D. and Vyrastekova, J. (2004), Peer Enforcement by Selective Exclusion Working Paper, Department of Economics: Tilburg University.

Taylor, L. (1987), 'The River Would Run Red with Blood: Community and Common Property in Irish Fishing Settlement', in B. McCay and J. Acheson, eds., *The Question of the Commons: The Culture and Ecology of Communal Resources* (pp. 290–307). Tuscon: University of Arizona Press.

Young, H. P. (1998), *Individual Strategy and Social Structure: An Evolutionary Theory of Institutions*, Princeton: University Press.

Appendix A. Proofs

Proof Lemma 1

(i) Since $Z(nx_H) > 0$, it follows that $\gamma > \delta$ implies $x_H Z(X) - \delta > x_L Z(X) - \gamma$ for all $X \leqslant nx_H$. Hence $\pi_1^D > \pi_1^E$ and, a forteriori, $\pi_2^D > \pi_2^E$. If $(x_H - x_L)Z(nx_L) < 2\delta - \gamma$ then $x_H Z(X) - 2\delta < x_L Z(X) - \gamma < 0$ for all $X \leqslant nx_H$ and hence $\pi_1^E > \pi_2^D$. If $\gamma > \delta$ and $(x_H - x_L)Z(nx_L) < 2\delta - \gamma$ then $x_H Z(X) - \delta < x_L Z(X)$ for all $X \leqslant nx_H$ and hence $\pi_0^E > \pi_1^D$.

(ii) If $(x_H - x_L)Z(nx_L) > \delta - \frac{1}{2}\gamma$ then $x_H Z(X) - \delta > x_L Z(X) - \frac{1}{2}\gamma < 0$ for all $X \leqslant nx_H$ implying $\pi_1^D > \frac{1}{2}(\pi_0^E + \pi_1^E)$. Moreover, the sanction rate is relatively low.

(iii) If $(x_H - x_L)Z(nx_L) < \delta - \frac{1}{2}\gamma$ then $x_H Z(X) - \delta < x_L Z(X) - \frac{1}{2}\gamma < 0$ for all $X \leqslant nx_H$ implying $\pi_1^D < \frac{1}{2}(\pi_0^E + \pi_1^E)$. A forteriori $(x_H - x_L)Z(X) < 2\delta - \gamma$ for all $X \leqslant nx_H$, so that the sanction rate is relatively low.

(iv) If $(x_H - x_L)Z(nx_L) < \delta - \gamma$ then $x_H Z(X) - \delta < x_L Z(X) - \gamma$ for all $X \leqslant nx_H$, implying $\pi_1^E > \pi_1^D$. Then also $\pi_2^E > \pi_2^D$ because $\delta > \gamma$. If $\delta - 2\gamma < (x_H - x_L)Z(nx_H)$ then $0 < x_H Z(X) - \delta > x_L Z(X) - 2\gamma$ for all $X \leqslant nx_H$, implying $\pi_1^D > \pi_2^E$.

Proof Proposition 2

(i) Number the positions on the circle clockwise. Put an enforcer on position 1 and, without loss of generality, a defector on position 2. Suppose $n = 2$. This is not an equilibrium because $\pi_1^D > \pi_1^E$. Suppose $n = 3$. This case is ruled out by lemma 3(iii) or lemma 3(iv). Suppose $n = 4$. At number 3 there is a defector in view of lemma 3(iv). At number 4 there is an enforcer in view of lemma 3(iii). But this cannot be an equilibrium because $\pi_1^D > \pi_1^E$. Suppose $n = 5$. At number 3 there is a defector in view of lemma 3(iv). At number 4 there is an enforcer in view of lemma 3(iii). At number 5 there is an enforcer because of lemma 3(iv). So the equilibrium candidate looks like: EDDEE. This is indeed an equilibrium. The defectors will remain defectors since $\pi_1^D > \pi_1^E$ and the enforcers will remain enforcers since $\pi_1^D < \frac{1}{2}(\pi_0^E + \pi_1^E)$. Next we show that the minimal length of an E-cluster is equal to three. Suppose there exists a DE equilibrium (with $n \geqslant 5$) with only two adjacent enforcers, surrounded by defectors: DEED. Then, because of lemma 3 we must also have DEEDD. This cannot be (part of) an equilibrium because $\pi_1^D > \pi_1^E$.

(ii) Consider a CDE configuration. Put the cooperator closest to a defector on position 1. Suppose the first defector is at number 2. This contradicts lemma 3(ii). Suppose the first defector is at number 3. There is an enforcer at number 2 by construction. This cannot be an equilibrium in view of lemma 3(i). Suppose the first defector is at number 4. There are enforcers at numbers 2 and 3 by construction. This cannot be an equilibrium because the enforcer at number 2 will turn into a cooperator since $\pi^C > \frac{1}{2}(\pi_0^E + \pi_1^E)$. Suppose the first defector is at number 5. At numbers 2, 3 and 4 there are enforcers by construction. There cannot be an enforcer at number 6 because of lemma 3(iv). There cannot be a cooperator at number 6 by construction. Hence is a defector at number 6. Because of symmetry there are enforcers at numbers 7, 8, and 9. It is easily verified that this is an equilibrium. Therefore the minimal number of players necessary for a CDE equilibrium is 9. Suppose there is a CDE equilibrium with a string ED. We cannot have CED in view of lemma 3(i), nor DED (lemma 3(iii)). So, we have a string EED. We cannot have DEED by the following reasoning. If the further extension could be written as DEEDD then this cannot be an equilibrium because $\pi_1^D > \pi_1^E$, implying that the second enforcer in the row turns into a defector. Lemma 3(ii) rules out the further extension DEEDC. And the extension DEEDE is not allowed in view of lemma 3(iv). Therefore, DEED cannot be part of an equilibrium. Consider, therefore, CEED. Again the further extension cannot be CEEDD, CEEDC or CEEDE. Hence we should have EEED. Therefore, the minimal string of enforcers is 3 if an enforcer is adjacent to a defector.

(iii) In a blinker an enforcer will never switch to defection, for the following reason. An enforcer next to a defector will switch to defection only if it punishes twice: with CED the enforcer switches to cooperation, and with EED the (second) enforcer stays an enforcer since $\pi_2^D < \pi_1^D < \frac{1}{2}(\pi_0^E + \pi_1^E)$. Therefore, we must have DED. But the first defector will not switch to enforcement since $\pi_2^D > \pi_2^E$. It follows that DE blinkers do not exist. In a CDE blinker a cooperator will never switch to enforcement. Therefore, there should be a defector switching to enforcement. A necessary condition is that we have EDE. But the first enforcer will not switch to defection.

Appendix B. Average Frequencies of Equilibria for Different Sanction Levels

Table B.1. Average frequency of convergence for different sanction levels

sanction	D-equil.	DE-equil.	CE-equil.	CDE-equil.	CDE-equil. (blinking)	CDE-equil. (cycling)	E-equil.
140	1.00	0.00	0.00	0.00	0.00	0.00	0.00
160	1.00	0.00	0.00	0.00	0.00	0.00	0.00
180	1.00	0.00	0.00	0.00	0.00	0.00	0.00
200	0.99	0.00	0.01	0.00	0.00	0.00	0.00
220	0.99	0.00	0.01	0.00	0.00	0.00	0.00
225	0.99	0.00	0.01	0.00	0.00	0.00	0.00
230	0.99	0.00	0.01	0.00	0.00	0.00	0.00
235	0.60	0.34	0.01	0.05	0.00	0.00	0.00
240	0.40	0.39	0.01	0.20	0.00	0.00	0.00
245	0.34	0.36	0.01	0.29	0.00	0.01	0.00
250	0.32	0.34	0.03	0.28	0.00	0.03	0.00
260	0.32	0.34	0.03	0.28	0.00	0.03	0.00
270	0.32	0.35	0.03	0.28	0.00	0.03	0.00
280	0.32	0.33	0.04	0.29	0.00	0.02	0.01
290	0.32	0.33	0.04	0.29	0.00	0.02	0.01
300	0.32	0.33	0.04	0.29	0.00	0.02	0.01
310	0.31	0.33	0.04	0.29	0.00	0.02	0.01
320	0.32	0.33	0.04	0.29	0.00	0.02	0.01
350	0.30	0.34	0.04	0.29	0.00	0.02	0.01
360	0.30	0.33	0.04	0.30	0.00	0.02	0.01
370	0.30	0.34	0.04	0.30	0.00	0.02	0.01
500	0.12	0.34	0.06	0.45	0.01	0.01	0.01
700	0.10	0.44	0.07	0.33	0.02	0.01	0.02
750	0.09	0.45	0.07	0.33	0.03	0.01	0.02
800	0.00	0.01	0.78	0.00	0.01	0.00	0.20
900	0.00	0.00	0.78	0.00	0.01	0.00	0.21

10. Sustainable Motion in Classical Mechanics: An Economics Perspective

JOHN M. HARTWICK

Queen's University, Kingston, Ontario, K7L 3N6 Canada (e-mail: hartwick@econ.queensu.ca)

1. Introduction

A central organizing principle in economics is that per period the value of product or output "factors back" and thus equals the value of inputs. Presumably this idea got firmly rooted with Walras's Elements, of 1874. Here we draw on Hamilton's formulation of classical mechanics to tease out the same sort of account, in units of energy rather than dollars, for the motion of "a particle" in classical mechanics (CM). At each instant of time, the energy value of output (a vector comprising position-change for a particle and velocity or momentum-change) is balanced with the energy value of input flow from capital, also a vector with two components. Sustainable or periodic activity has the capital goods restored by investment to their initial values over the period in question. The Virial Theorem in CM is an energy account which reflects this restoration of position variables over the period and we indicate that there is a corresponding restoration account in units of energy for the momentum variable over the period. Planetary motion (Kepler's model developed by Newton) is of course one striking instance of periodic activity in CM. Motion of the undamped pendulum is another well-known case.

The restoration of state variables or capital values over a period is an explicit basis for the idea of sustainable activity. Non-sustainable activity in CM involves capital values never returning to their "initial" values and corresponding to this is inherent cross-subsidization of either position-change for a particle in motion or momentum-change by one type of capital or the other. There is then in CM an unambiguous notion of capital being maintained intact for the case of periodic or sustainable activity. Recall that in the 1974 Solow model of sustainable consumption, output $K^{\alpha} R^{\beta}$ is unchanging and rentals from capital, KF_K and RF_R are also unchanging. Capital goods do not get restored to their "initial" values, however.

We proceed to set out the Hamilton equations for equilibrium motion in CM and to derive the energy account balance of inputs with outputs at each instant. We then consider the special case of sustainable motion (the Virial Theorem and the Complementary Virial Theorem) and illustrative examples. We close with some more general remarks.

L. Bretschger and S. Smulders (eds.), Sustainable Resource Use and Economic Dynamics, 217–227.

2. Energy Accounting for Particle Motion in Classical Mechanics

In brief, our argument is the following. Particle motion satisfies the Hamilton equations

$$\frac{\partial H}{\partial q_i} = -\dot{p}_i$$

and

$$\frac{\partial H}{\partial p_i} = \dot{q}_i, \quad i = 1, ..., 3$$

for H the Hamiltonian, q_i a component of a vector of location (state) variables, and p_i a component of a vector of momentum (co-state) variables; i here indicates three spatial coordinates. We are comfortable with reading the Hamilton equations as: first, net force is causing current momentum-change and second, velocity is causing current position-change for the particle. Lemons (2002; p. 53), for example, speaks of particle motion characterized by "force" causing "change of velocity", a particular case of the first equation, and "velocity" causing "change of position", a particular version of the second equation.[1] We multiply the first by q_i and the second by p_i to get equations in units of energy.

$$q_i \frac{\partial H}{\partial q_i} = -q_i \dot{p}_i$$

and

$$p_i \frac{\partial H}{\partial p_i} = p_i \dot{q}_i$$

$q_i \partial H/\partial q_i$ is "generalized net work" in units of energy.[2] It is a measure of the net effect of force on the particle in its current location. (We illustrate with examples below.) We interpret this "generalized work" as a measure or current energy inflow, attributable to location, on the particle in motion. $p_i \partial H/\partial p_i$ is an energy measure of the effect of the particle's momentum on itself. We interpret this as a measure in units of energy of the particle's current momentum on its motion.[3] Again we see this as an energy measure of current energy inflow to current particle motion. The right-hand sides are respective measures of energy useflow associated respectively, with momentum-change for the particle and position-change for the particle. A verbal rendering of these equations is: first, net force, in units of energy is causing current momentum-change, in units of energy, and second, momentum, in units of energy is causing current particle position-change, measured in units of energy.

Our aggregate capital income-expenditure balance, per instant, energy equation is then

$$\sum_{i=1}^{3} \left\{ \left[p_i \dot{q}_i - q_i \frac{\partial H}{\partial q_i} \right] + \left[-q_i \dot{p}_i - p_i \frac{\partial H}{\partial p_i} \right] \right\} = 0 \tag{1}$$

where q_i is a capital good in the first bracket and p_i is a capital good in the second bracket. $\dot{q}_i$ and $\dot{p}_i$ operate as respective investment terms. $\partial H/\partial q_i$ and $\partial H/\partial p_i$ are operating as respective rental prices, which translate "quantities" into units of energy. p_i in the first bracket is a capital goods price which translates investment, $\dot{q}_i$ into units of energy and q_i in the second bracket is a capital goods price which does the analogous thing for investment, $\dot{p}_i$.

3. Momenta as Capital Inputs

Our treatment of momentum variables as capital goods might strike an economist as odd. One can obtain Hamilton's equations from a dynamic optimization problem as "action sum" minimization over a finite interval.[4] In such a problem, the location variables are natural state variables and the momentum variables are natural co-state variables. Hence an economist might see it as natural to interpret location variables as capital goods and co-state variables as capital goods shadow prices. This goes through well in the interpretation of the Virial Theorem as a statement of energy inflow balanced with energy useflow, per period. But in physics, the co-state variables capture much more than being prices or "translators" of quantities into units of energy. For example, when a simple pendulum swings through its low point, there is no external force acting to move the bob. All force is acting to create tension on the rod holding the bob and yet the bob (the particle in motion) is moving most rapidly and is exhibiting most kinetic energy. The answer to this paradox is of course that it is the momentum of the bob that is causing the "large" motion at this low point. This momentum "driving" the bob has an energy representation analogous to force having an energy representation. Energy inflow from current momentum is captured above in the term $p_i \partial H/\partial p_i$. This energy inflow is the analogue of net generalized work, namely $q_i \partial H/\partial q_i$. One might refer to the latter as the energy inflow associated with local net force. We see a large symmetry here with momentum having an energy inflow representation and location, standing in for local force, having an energy inflow representation. This may be our central departure from textbook physics, namely the symmetric treatment of location variables and momentum variables in units of energy.

One can arrange things so that location variables are always non-negative. This is not true for momentum variables. Hence we have the anomaly of a capital input with a negative sign over some intervals of time. Kinetic energy, half the current value of investment expenditure on location change, is always non-negative. But other prices and quantities can be negative. This makes energy accounting in classical mechanics different from value accounting in economic dynamics. Physics can be said to be more general, in this regard. The bottom line of energy accounting for us is the fact that for the case of periodic motion, we observe a positive inflow of energy per period from both location variables and momentum variables and these flows balance with positive use-flows of energy per period by location-change and momentum-change, respectively.

4. Three Illustrations from Classical Mechanics

4.1. PARTICLE FREE-FALL

This classic Galilean problem in a locally unchanging gravitational field has potential energy, mgz for potential energy normalized to be zero when z is zero. m is particle mass, g is "gravitational acceleration", and z is a "vertical" coordinate. The particle falls from some initial positive value of z toward z equal to zero. Velocity v is negative here, with $v = \dot{z}$. Kinetic energy is $(1/2)mv^2$. The Hamiltonian is $H = mgz - (1/2)mv^2 + pv$, for p the co-state or momentum variable. p gets defined in this formulation by $\partial H/\partial v = 0$ or $p = mv$, with p here also negative. There is only one location variable here. The Hamilton equations are

$$mg = -\dot{p} \quad \text{and} \quad v = \dot{z}.$$

In units of energy, we have

$$zmg = -z\dot{p} \quad \text{and} \quad pv = p\dot{z}.$$

Net investment here is

$$\{p\dot{z} - zmg\} + \{-z\dot{p} - pv\}$$

which is zero at each instant of motion. The energy associated with current changes in location and momentum (investment expenditure, in units of energy), namely $p\dot{z} - z\dot{p}$, is supplied by "capital rentals" associated with z and p, namely $zmg - pv$.

4.2. THE SIMPLE UNDAMPED PENDULUM

In this case of periodic motion, there is a single location (state) variable, θ, the pendulum rod angle with respect to the vertical "origin". p is the single co-state variable. Potential energy is mgh for $h = [1 - \cos\theta]$. This normalizes the potential energy function to have value zero when the particle or bob swings through its low point. Kinetic energy is $(1/2)mgv^2$ for $v = l\dot{\theta}$, $l$being the length of the pendulum rod holding the bob. The Hamiltonian is $H = mg[1 - \cos\theta] - (1/2)mgv^2 + pv$. Here, p gets defined by $\partial H/\partial v = 0$ or $p = mv$. The Hamilton equations are

$$\frac{\partial H}{\partial \theta} = -\dot{p} \quad \text{and} \quad \frac{\partial H}{\partial p} = \dot{\theta}$$

and in units of energy are[5]

$$\theta\frac{\partial H}{\partial \theta} = -\theta\dot{p} \quad \text{and} \quad p\frac{\partial H}{\partial p} = p\dot{\theta}.$$

At an instant, net investment is

$$\left\{p\dot{\theta} - \theta\frac{\partial H}{\partial \theta}\right\} + \left\{-\theta\dot{p} - p\frac{\partial H}{\partial p}\right\}.$$

which is zero. Net investment expenditure on current position-change, namely $\left\{p\dot{\theta} - \theta \partial H/\partial \theta\right\}$, plus net investment expenditure on current momentum-change, namely $\{-\theta \dot{p} - p\partial H/\partial p\}$, is zero.

The Virial Theorem (Goldstein et al. 2002) for this problem[6] is,

$$\int_0^B \left\{p\dot{\theta} - \theta \frac{\partial H}{\partial \theta}\right\} dt = 0$$

for B the period of motion.[7] One also has

$$\int_0^B \left\{\theta \dot{p} - p \frac{\partial H}{\partial p}\right\} dt = 0$$

which we refer to as the complementary virial theorem, a zero energy relationship, per period, for the momentum variable. Each equation has the interpretation of investment expenditure per period, balancing with own capital rentals per period. The sum over the period of energy inflow and useflow is POSITIVE in each subaccount above. This is an accounting of particular energy inflows sustaining distinctive OWN motion over the period: position-change and momentum-change. This was the central message of our earlier paper.

$\partial H/\partial \theta$ is $\partial U/\partial \theta$ or force here and $\partial H/\partial p$ is a measure of velocity of the particle here. Hence energy inflow reduces to $\theta \partial U/\partial \theta$ in the first equation. $\theta \partial U/\partial \theta$ is "generalized work" as distinct from $U(\theta)$, potential energy. Hence "generalized work" becomes a central concept in our formulation of energy accounting for classical mechanics.[8] We have a form of energy conservation per period (actually an account of energy inflow balanced with energy useflow per period) without potential energy in the accounting.

In addition to satisfying these "sustained motion" equations immediately above, one has capital income–expenditure balance per instant, in the sense of equation (1). Particle motion is generally then: capital income sustaining current expenditure on capital investment and disinvestment. The general case requires a summing over different types of capital. For the special case of periodic motion, each type of capital stands on its own bottom, so to speak; there is no cross-subsidization of investment in either location or momentum.

4.3. THE CLOSED ORBIT KEPLER PROBLEM

The Hamilton equations specialize to

$$\begin{aligned} \frac{dU}{dr} - \frac{dT}{dr} &= -\dot{p}_r \\ 0 &= \dot{p}_\theta \\ v_r &= \dot{r} \\ v_\theta &= \dot{\theta} \end{aligned}$$

for $H = U(r) - T + p_r v_r + p_\theta v_\theta$, $U(r) = -\chi/r$,[9] $T = (1/2)mv_r^2 + (1/2)mr^2 v_\theta^2$, $v_r = \dot{r}$, and $v_\theta = \dot{\theta}$. Here $U(r)$ is potential energy, T is kinetic energy, m is mass,

v_r and v_θ are location coordinates for the particle in motion, and p_r and p_θ are momentum variables.

The Virial Theorem can be expressed as $\int_0^B r\,\{dU/dr - dT/dr\}dt = \int_0^B \dot{r}\,p_r dt$ for B the period of orbital motion.[10] This can be rewritten as $\int_0^B r\,\{dU/dr\}dt = \int_0^B \left\{\dot{r}\,p_r + \dot{\theta}\,p_\theta\right\}dt$ which has the interpretation of location capital rentals, $r\{dU/dr\}$ funding investments in location variables.[11] The complementary virial theorem is $\int_0^B p_r v_r dt = \int_0^B \{-r\dot{p}_r\}\,dt$, which as the interpretation of rentals associated with capital good p_r funding investment in p_r, over the period of motion. Capital good p_θ supplies energy but does not need "investing in" because its value is unchanging over the period. Capital good θ supplies no energy but must be invested in, in order to restore it to its initial value over the period of motion. Periodic motion is motion involving "maintaining capital goods" over the period.

Our generalization of these arguments to an instantaneous energy account of capital income–expenditure balance involves taking the "energy version" of the Hamilton equations and expressing them as

$$\left\{\dot{r}\,p_r - \left[\frac{dU}{dr} - \frac{dT}{dr}\right]r\right\} + \left\{\dot{\theta}\,p_\theta\right\} + \{r\,\dot{p}_r - p_r v_r\} + \{-p_\theta v_\theta\} = 0.$$

The integral over the period of the first term is zero (the Virial Theorem) and the integral over the period of the third term is zero (the complementary virial theorem), and the second and fourth terms sum to zero, per instant.

5. The Conserved Quantity

One of the best known results in classical mechanics is "conservation of energy" under equilibrium motion. One way to view this is to write down the Hamiltonian for a problem in conservative motion, as in $H(t) = U(q(t)) - 0.5v(t)^2 + p(t)v(t)$, and then write down the derivative, $\dot{H}(t)$and observe that the Hamilton equations of equilibrium motion imply $\dot{H}(t) = 0$. This is a proof of: equilibrium motion implies conservation of energy, where $H(t)$ is the so-called current total energy (the sum of current kinetic and potential energies) for the problem under consideration. Our result for periodic or aperiodic motion can be stated: equilibrium motion implies that current energy inflow (value of inputs in units of energy) balances with current energy use-flow (investment expenditures) at each instant, i.e. $[q\partial H/\partial q - p\dot{q}] + [p\partial H/\partial p + \dot{p}q] = 0$. Alternatively we have current energy inflow value "conserved" in current energy useflow, where the flows are associated with both momenta and location variables. Novel here is the concept of current sum, $S(t)$ satisfying

$$\dot{S}(t) = \left[q\frac{\partial H}{\partial q} - p\dot{q}\right] + \left[p\frac{\partial H}{\partial p} + \dot{p}q\right].$$

Since the Hamilton equations imply that the right-hand side is always zero, we could say that equilibrium motion implies that sum $S(t)$ is conserved. In this view,

$[q\partial H/\partial q - p\dot{q}]$ is then marginal sum, in units of energy, associated with location variables and $[p\partial H/\partial p + \dot{p}q]$ is marginal sum associated with momentum variables. Sum, here, would then be a new quantity conserved by equilibrium motion. In accord with Noether's Theorem (Rasband, 1983; pp 136-39), there is usually a symmetry associated with a conservation property. A symmetry usually takes the form of the Lagrangian of a system being invariant to a parameter change. We have no conjecture on the form of a possible symmetry associated with our $\dot{S}(t) = 0$. However, we suggest that it is appropriate to view "income equal to expenditure" at each instant as the outcome of a current surplus maximization operation. The operation we have in mind is completely analogous to surplus maximization in economics, with the outcome there being commodity price equalling its marginal cost. Such an outcome is obviously closely related to marginal expenditure on a commodity being equal to marginal income derived from the production of that commodity.

Generalized work on the input side has an energy value the same as "product value" at each instant. The notion of "generalized work" must be expanded to include energy inflows from current momentum values as well as current location values, where these latter capture the values of local forces. "Product value" is the energy representation of current particle position-change and momenta-changes. A system with friction, a non-conservative system, will not possess this value–balance relation since some energy input will not show up in current position-change and momenta-changes. Some energy inflow will be lost to heat or the manifestation of the friction.

Our energy account and view of what is going on with particle-motion in classical mechanics is very different from the standard textbook view in physics. The standard view sees the current diminution in potential energy as "input" and the current increase in the particle's kinetic energy as "output". This is "the work-energy theorem" and was set out by Newton in Proposition 39, Book 1 of his *Principia*. He was not focusing on energy relations there but did produce "the work-energy theorem" in his analysis of particle free fall in a general force field. Though "the work-energy theorem" and the related "conservation of energy" has served physics extremely well since Newton, it is not a conceptual scheme that connects well to ways of conceptualizing in say economics. Once one asks about the nature of current "product" in classical mechanics and current "input", one is drawn back to the drawing board to, we suggest here, an account such as we have developed. It may not be an account that yields new physics but it allows for a different conceptual framework for what is going on in classical mechanics.

The great and iconclastic Feynman suggested somewhat indirectly that there is a need for a reconceptualization of what the equations of classical mechanics are saying. Consider this passage from Feynman's well-known lectures. This passage follows his own explanation or reporting of particle motion on a closed Kepler orbit.

"What is gravity? ... All we have done is to describe how the earth moves around the sun, but we have not said **what makes it go**. Newton made no hypotheses about this; he was satisfied to find what it did without getting into the machinery of it. **No one has since given any machinery**. It is characteristic of physical laws that they have this abstract character. The law of conservation of energy is a theorem concerning quantities that have to be calculated and added together, with no mention of the machinery, and likewise the great laws of mechanics are quantitative mathematical laws for which no machinery is available. Why can we use mathematics to describe nature without a mechanism behind it? No one knows." (Feynman, Leighton, and Sands 1963, pp. 7–9, their emphasis)

Feynman is actually making a number of different observations here but he is explicit in asserting that conservation of energy is not intuitive. He also seems to be saying that the detailed equations of motion do not pay the effort at intuiting. Our view is that the reason the equations are counter-intuitive is that force and momentum are generally scrambled together in the equations. In a sense the Hamilton formalism allows for the separation of the roles of location and momentum in particle motion in classical mechanics. This is of course not Feynman's explicit observation here. He is not rejecting the view that conservation of energy provides a valuable benchmark for organizing thinking about problems in physics but the law itself possesses no compelling intuition, directly. Indirectly however, conservation of energy is linked to the "time symmetry" or "time autonomousness" of the laws of physics.

A further critique of physics textbooks. The Virial Theorem for periodic motion has twice the sum of kinetic energy over the period as the implicit measure of "ouput". And it has the sum of "generalized work" over the period as the implicit measure of "input". This to us is fine but we would like to see the labels' energy inflow over the period and energy "consumption" or use-flow over the period attached to the two sides of the energy equation. What is missing is what we call "the complementary virial theorem" dealing with energy inflow linked to momentum variables and energy useflows linked to momentum-changes over the period. The energy representation of a current momentum, $p_i \partial H/\partial q_i$, requires a label, a counterpart to the generalized work associated with force. And the energy representation of current momentum-change, namely $-\dot{p}_i q_i$, also requires a label. Kinetic energy is the term used for energy associated with current particle position-change. We are not asking for new physics, here. We are asking for new labels of terms in a new energy account, an account built around the idea of current "product" factoring back into current "input".

Newton created modern classical mechanics from Galileo's treatment of local, particle free fall and Kepler's three laws. In his own words, Newton created a mathematical theory of force, both local and celestial. Newton very explicitly asked that his theory generate observed trajectories of particle motion. He said (we paraphrase) on many occasions in the *Principia*: "Proposition i involves solving for the position of a particle and the interval of its motion, relative to its initial position". He then solved for equations generating trajectories and time lapses of motion. Never did he say "Proposition i involves solving for the trajectory and the corresponding path of velocity-changes of a particle and its interval of motion, relative to its initial position and velocity". Such a formulation would have emphasized our point about particle-motion being a production system, with current "product" being defined by current particle position-change and velocity-change.

Newton was not wrong. He seems to have been intellectually cautious. He was able to analyse motion to his satisfaction with his mathematical theory or force but it seems that he was aware that force was a fairly controversial entity. He chose not to digress into a larger theory of motion involving energy accounting or "input" and "output" detail. There is a sense in which the *Principia* contains the bare-bones theory of particle motion and that suited Newton's purposes. Significant "extensions" were made by Euler, Lagrange, and Hamilton. Hamilton's view of particle motion, implicit in "the Hamilton equations" is, in our view, a significant refinement of Newton's view since it treats particle motion as "equally weighted", simultaneous position-change and velocity-change. Newton focused his analysis of particle motion on position-change and the interval of motion. But Hamilton's view has not displaced Newton's view in college-level textbooks. In Hartwick (2004) we presented an energy accounting based on Hamilton's view for periodic motion and here we re-present that accounting, with extension to general particle motion, not simply periodic motion. We did not start out to construct an energy account for particle motion based on the Hamilton equations but our inquiry about current "product" and current "input" in classical mechanics led us to develop such an account.

6. Concluding Remarks

We have been motivated to investigate the nature of motion in classical mechanics from the standpoint of economic accounting. One has a system with a product, namely particle motion, and one is motivated to factor back output into inputs. And one has the large question of current value balance: the value of current output being reducible to the value of inputs. We have succeeded in a reconceptualization of what is going on in particle motion in classical mechanics. The small bit of new physics we have obtained from our alternative view is that per period energy inflow and useflow balance suggests a natural normalization of the potential function. Perhaps something more substantive will emerge in the future.

The starting point of our analysis is the view that at each instant of time, particle motion in classical mechanics is a system with a "product" and "inputs". We have observed that this intuition can be made concrete, particularly by drawing upon economics ideas. The "output" associated with particle motion is in general a simultaneous bit of position-change for the particle and momenta-changes for the particle. In units of energy these outputs can be "factored back" into values of inputs, at a moment in time. Twice current kinetic energy is the energy measure of current particle position change, and $-\dot{p}_i q_i$ is the energy value of current change in momentum i.

The Virial Theorem in classical mechanics for periodic and quasi-periodic motion provides a concrete case of "input value" (rental income from capital inputs) in units of energy balanced with "product value" (investment expenditure) in units of energy per period. However the Virial Theorem deals only with the energy balance of position variables.

There is a complementary virial theorem for periodic motion dealing with energy inflow per period associated with momentum variables balanced with energy useflows associated with change-in-momentum values. This balance relation holds for momentum variables in a precisely analogous way as does the Virial Theorem for location variables.

For non-periodic motion, we observe at each instant of time a balance of the energy value of "product" with the energy value of "input", with "product" comprising the sum of the energy values of current position-change and momenta-changes for a particle in motion.

Position-change and momenta-changes in their energy representations have the interpretation of current investment expenditure in economics. The energy inflow associated with force or with the position of the particle has the interpretation of capital good incomes or rental flows. And the analogous energy inflow associated with momenta variables has the interpretation of capital good incomes or rental flows. The main instantaneous energy account in classical mechanics reads: the sum of capital input rentals at an instant balances with investment expenditures on capital, all in units of energy.

Notes

1. Newton's second law identifies the net force $F(t)$ per unit particle mass, M with the rate at which the particle changes its velocity V(t). This velocity, in turn, describes the rate at which the particle changes its position.
2. Work is force multiplied by distance and is in units of energy. Work usually refers to energy involved in the horizontal movement of a mass. By generalized work, we mean "force multiplied by distance" in an abstract setting.
3. We like the interpretation of generalized work causing momentum change, in units of energy, and momentum, in units of energy, causing particle position-change. Our approach may part with standard textbook classical mechanics most when we define and make use on an energy representation of current momentum. But we also depart from conventional energy accounting by substituting generalized work for potential energy, roughly speaking.
4. Hamilton's principle is that our "action sum" is extremized by a path of equilibrium motion of a particle. We label action, current potential energy minus current kinetic energy. This appears not standard. More standard is referring to our "action sum" as "action".
5. Generalized work here is $\theta\partial H/\partial\theta$. Our analysis turns on the symmetric object, also in units of energy, namely $p\partial H/\partial p$.
6. The Virial Theorem is usually expressed as a time-average or in energy per instant rather than energy per period.
7. We suggest that the natural normalization of the potential energy function for this problem yields $\int_0^B U(\theta)dt = \int_0^B \theta U_\theta(\theta)dt$.
8. In our earlier paper we emphasized that one could always make $\int_0^B \theta(\partial U/\partial\theta)dt$ equal to $\int_0^B U(\theta)dt$ by an appropriate normalization of $U(\theta)$.

9. Potential energy is negative for this problem. Convention has the potential energy function normalized so that potential energy tends to zero, far from the central force.
10. Here generalized work is $r\{dU/dr\}$, distance multiplied by force.
11. $r\{dU/dr\} = -U(r)$ under the traditional normalization of the potential function. We identify $\int_0^B -U(r)dt$, with total energy associated with the action of central force on the particle in motion, per period.

References

Feynman, R. P., Leighton, R. B., and M. L. Sands (1963), *Lectures on Physics*, Vol 1. Reading, MA: Addison-Wesley.

Goldstein, Herbert, C. Poole, and J. Safko (2002), *Classical Mechanics*. 3rd edn. New York: Addison-Wesley, p. 86.

Hartwick, J. M. (2004), 'Sustaining Periodic Motion and Maintaining Capital in Classical Mechanics', *Japan and the World Economy* **16**: 337–358.

Lemons, D. S. (2002), *An Introduction to Stochastic Processes in Physics*. Baltimore, MD: Johns Hopkins University Press.

Rasband, S. N. (1983), *Dynamics*. New York: Wiley.

Solow, R. M. (1974), 'Intergenerational equity and Exhaustible Resources' *Review of Economic Studies* Symposium: 29–46.

Index

The Economics of Non-Market Goods and Resources

1. John Loomis and Gloria Helfand: *Environmental Policy Analysis for Decision Making*. 2001 ISBN 0-7923-6500-3
2. Linda Fernandez and Richard T. Carson (eds.): *Both Sides of the Border*. 2002 ISBN 0-4020-7126-4
3. Patricia A. Champ, Kevin J. Boyle and Thomas C. Brown (eds.): *A Primer on Nonmarket Valuation*. 2003 ISBN 0-7923-6498-8; Pb: 1-4020-1445-7
4. Partha Dasgupta and Karl-Göran Mäler (eds.): *The Economics of Non-Convex Eco-systems*. 2004 ISBN 1-4020-1945-9; Pb: 1-4020-1864-9
5. R.T. Carson, M.B. Conaway, W.M. Hanemann, J.A. Krosnick, R.C. Mitchell, and S . Presser: *Valuing Oil Spill Prevention: A Case Study of California's Central Coast*. 2004 ISBN: 0-7923-6497-X
6. Riccardo Scarpa and Anna A. Alberini (eds.):*Applications of Simulation Methods in Environmental and Resource Economics*. 2005 ISBN: 1-4020-3683-3
7. Nancy E. Bockstael and Kenneth E. McConnell: *Environmental and Resource Valuation with Revealed Preferences*. 2007 ISBN 0-7923-6501-1
8. Barbara J. Kanninen (ed.): *Valuing Environmental Amenities Using Choice Experiments*. 2007 ISBN 1-4020-4064-4
9. Ståle Navrud and Richard Ready (eds.): *Environmental Value Transfer*. 2007 ISBN 1-4020-4081-4
10. Lucas Bretschger and Sjak Smulders (eds.): *Sustainable Resource Use and Economic Dynamics*. 2007 ISBN 978-1-4020-6292-6

Kolibri 360 GmbH
Ferdinand-Jühlke-Straße 7
99095 Erfurt, Deutschland
produktsicherheit@kolibri360.de